abc's of Hi-Fi and Stereo

by Hans Fantel

HOWARD W. SAMS & CO., INC.
THE BOBBS-MERRILL CO., INC.
INDIANAPOLIS · KANSAS CITY · NEW YORK

THIRD EDITION

FIRST PRINTING—1974

Copyright © 1963, 1967, and 1974 by Howard W. Sams & Co., Inc., Indianapolis, Indiana 46268. Printed in the United States of America.

All rights reserved. Reproduction or use, without express permisison, of editorial or pictorial content, in any manner, is prohibited. No patent liability is assumed with respect to the use of the information contained herein.

International Standard Book Number 0-672-21044-4
Library of Congress Catalog Card Number: 74-78351

Preface

The 20th-century music box has come a long way from its tinkling Swiss ancestors. Instead of mechanical imitations of music, modern sound reproduction conveys the sound of an actual performance.

Sound reproduction can be thought of as a means of transferring music through time and space—in a sense it delivers music to your home. But if you have low-quality equipment, the music arrives damaged. True, the tune always remains recognizable, and to some people that is all that matters. But there are others who like to hear music that still has a "breath of life." These people realize that true tone is a part of musical meaning, and they enjoy the subtle coloration of the various instruments. It is for these more demanding listeners that high fidelity was developed.

When high fidelity is augmented by stereo or four-channel sound, musical realism goes a step farther. Not only is the sound truthfully transmitted, but the exact position of each player and singer can be sensed. The origin of each sound is precisely located in space so that the whole acoustic aura of the original performance is transplanted to your living room.

This book is written to put you on more familiar terms with audio and hi-fi equipment and the underlying theory. It is, of course, perfectly possible to enjoy high fidelity without knowing anything about its technical aspects. After all, you don't have to read cookbooks to enjoy a good meal; however, some understanding of the cooking process does increase your appreciation of the meal. And so it is with audio.

On the practical side, knowing how your sound equipment works, you can coax it into giving you the best performance of which it is capable. In short, a little technical understanding can help you get the most for your money.

HANS FANTEL

Contents

CHAPTER 1
SOUND—THE RAW MATERIAL OF AUDIO 7
    Pitch—Tone Color—Loudness—Sound and Electricity

CHAPTER 2
WHAT IS STEREO? 12
    Stereo Reproduction

CHAPTER 3
FOUR-CHANNEL SOUND 16
    Cost Factor—Quad Program Sources—Quad Equipment

CHAPTER 4
CONSOLES OR COMPONENTS 21
    Consoles—Component Systems—Comparison of Systems—Compact Systems

CHAPTER 5
ANATOMY OF A SOUND SYSTEM 27
    Turntable—Tone Arm—Record Changer—Cartridge—Amplifier and Preamplifier—Speaker—Tuner—Receiver—Tape Recorder

CHAPTER 6
QUALITY FACTORS IN COMPONENTS 34
> Amplifiers — Tuners — Speakers — Bantam Speakers — Phono Cartridges — Turntables — Tone Arms — Record Changers — Tape Recorders — Earphones

CHAPTER 7
SOUND VALUE FOR YOUR DOLLAR 76
> "Bottom Dollar" Stereo — "Golden Medium" Stereo — "Deluxe" Stereo — Four-Channel Budgeting — Where Extra Dollars Really Count — The Price of Power — A Strategy for Shopping

CHAPTER 8
KITS FOR CASH AND PLEASURE 86
> The Practically Foolproof Kit — Should You Buy or Build It?

CHAPTER 9
SETTING UP YOUR SOUND SYSTEM 90
> Speaker Hookup–Impedance Matching and Phasing — Turntable Adjustments — Input Level Controls — Acoustic Feedback — Ventilation

CHAPTER 10
GETTING THE MOST FROM YOUR SOUND SYSTEM . . . 97
> Controls — Room Acoustics and Your Favorite Chair — Keep It Clean — Tape Care — How Loud Should You Play It?

CHAPTER 11
FIRST AID 108
> Hum — Stylus Problems — Other Problems

INDEX 111

CHAPTER 1

Sound — The Raw Material of Audio

The first requirement of fidelity is something to be faithfully reproduced—in our case, the sound of music. It is the raw material as well as the end product of audio—the stuff that enters into the microphone and ultimately issues from the speaker.

Fortunately, you can comprehend the sound of music in concrete terms—it consists of vibrations of the air. When you play a fiddle, a trumpet, or any other instrument, you make the instrument vibrate. These vibrations impart pulses to the air surrounding the instrument, and these pulses spread like waves in water when you throw a stone into a pond. Your own voice is an instrument for setting up such waves (strictly speaking, pressure vibrations) in the air, and when these waves reach the ear they are perceived as sound.

The most astonishing aspect of sound is its limitless variety. High fidelity can perhaps be best defined as the attempt to reproduce accurately the greatest possible range of different sounds. Therefore, the factors that account for these differences in sound must be analyzed.

PITCH

When you hear a musical note, it seems either high or low; you notice its *pitch*. What you perceive as pitch is determined by the number of vibrations of the air within a given span of time; this is called the *frequency* of a tone, and each single vibration (or pulse) is called a *cycle*.

Are we suddenly getting technical? Not at all. The term cycle has exactly the same meaning here as in such familiar expressions as "weather cycle" or the "cycles" of the stock market. It simply means that the process repeats itself—that the vibrations follow each other at regular intervals and thereby create the pitch of a tone. The faster the vibration (i.e., the higher the frequency), the higher the tone seems to the ear (Fig. 1-1A).

All this can be summed up in what might be called the first principle of sound: Each musical note has a certain *pitch* de-

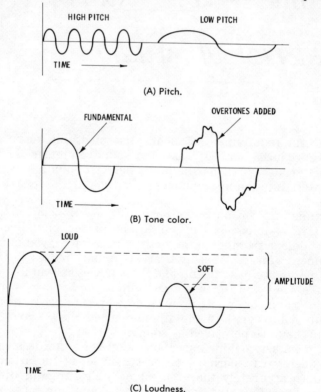

Fig. 1-1. Graphical representation of sound characteristics.

termined by its *frequency*, i.e., the number of vibrations (or cycles) in a given time. Frequency is expressed in *hertz* (abbreviated as *Hz*). One hertz equals one cycle per second.

There are some sounds without definite pitch—for example, splashing water, rushing air, jangling metal, and other kinds of noise. Such sounds represent a random mixture of frequencies; that is, the vibrations have no regular rate of recurrence.

The lowest sounds audible to the human ear (for instance, the roll of distant thunder) are around 16 Hz. The highest sounds humanly perceptible (hissing steam, metallic clinks) range up to 20,000 Hz. As a rule, only children and adolescents can hear these uppermost frequencies. In adults, hearing usually has an upper limit around 16,000 Hz and declines to 10,000 Hz or less as age increases.

TONE COLOR

An oboe and a violin may play the same note, but the sound they produce is different. What accounts for this?

In addition to its pitch frequency (called the *fundamental*), each instrument projects a whole series of other frequencies, which are called *overtones,* or *harmonics*. These are multiples of the fundamental frequency (i.e., the basic pitch frequency $\times$ 2, $\times$ 3, $\times$ 4, and so on). Not all these overtones are equally strong. Each instrument has its own pattern of emphasis on various overtones. This difference in relative emphasis on overtones makes instruments differ from each other in tonal color, or timbre, as it is technically known.

Now it becomes apparent why extended frequency range is needed for truthful sound. Suppose a flute is playing a 1,500-Hz note. The overtones would be 3,000 Hz (1,500 $\times$ 2), 4,500 Hz (1,500 $\times$ 3), 6,000 Hz, 7,500 Hz, 9,000 Hz, 10,500 Hz 12,000 Hz, and so on. The point is that these higher frequencies must be reproduced to bring out the instrumental color of the flute playing the 1,500-Hz note. It is in these overtones that we get the lifelike character of the instrument. Suppose a violin plays the same note. The overtones would be the same, but with a different pattern of emphasis. Even low-pitched instruments like the bass fiddle and the kettledrum generate high-frequency overtones that lend them their particular tonal flavor.

So put down as the second basic principle that tone color is produced by overtone patterns (Fig. 1-1B). To project the true character of an instrument, a high-fidelity system must therefore have a wide frequency range (at least to 15,000 Hz) to reproduce these overtones. The frequency ranges of various instruments and sounds are shown in Fig. 1-2.

LOUDNESS

Every sound also has a certain loudness, or volume, as it is loosely called. Suppose you knock at a door. If nobody answers, you bang harder. By this commonplace action you have demonstrated the physical difference between soft and loud sounds. Loud sounds are produced by a greater expenditure of energy, or, to put it the other way, louder sounds have greater energy content.

Since sound is motion of the air, try to imagine both pitch and loudness in terms of motion. Again visualize sound waves as similar to the waves on the surface of a pond. Waves spaced at long intervals (low frequency) would be low-pitched notes, short ripples would be high notes, and the height of each wave, which marks its energy content, would signify loudness. In the

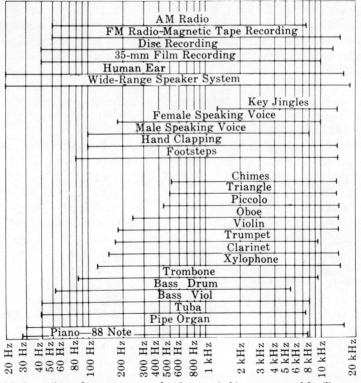

Fig. 1-2. Approximate frequency range of various musical instruments and familiar sounds.

language of physical measurement, the height of a wave is called its *amplitude*. As the third basic principle, therefore, you can say that the *amplitude* of a sound wave determines whether the sound is loud or soft (Fig. 1-1C).

SOUND AND ELECTRICITY

To many people the notion of music running through a wire must seem quite uncanny. The change of sound to electricity and vice versa is profoundly marvelous, even though it is a commonplace event that happens every time you pick up the phone. The relation of sound to a corresponding electric signal is, in fact, the key to the mysteries of audio.

To form a mental picture of what goes on, you may find it helpful to think of an electric current as a plastic conveyor belt moving at high speed. If you make this "belt" undulate exactly like the sound waves you want to hear, it then becomes a carrier of sound. In effect, the sound (or rather the waveform of the sound) rides piggy-back on the electric current.

This process is called *modulation*. When the electric current has been modulated, it assumes the frequencies and amplitudes of the music or speech you want to reproduce. Thus transformed into an electric *signal*, music and speech can be amplified, recorded, or sent out via radio. It is in this electric guise that music makes its journey through time and space into your living room.

CHAPTER 2

What Is Stereo?

Stereo was "invented" millions of years ago when two-eyed, two-eared creatures first appeared on earth. You perceive the world in "pairs" with your two eyes and two ears. That's what makes your sense of sight and sound three-dimensional.

STEREO EFFECT

The visual stereo effect is easily demonstrated. Look at pictures through a stereoscopic viewer and close one eye. Everything immediately goes flat. Or, for more exasperating proof of the importance of visual stereo, try threading a needle with one eye closed.

Something similar holds true of stereophonic hearing. It provides a kind of space perception not attainable by one-eared listening. You perceive depth and direction with your ears just as you do with your eyes. When you hear someone honking at you, you instinctively turn your head toward the approaching car. Blindfolded you can tell the approximate size of the room you're in by the sound reflection from its walls. Your own voice sounds different in the narrow confines of a phone booth or in the vaulted reaches of a great cathedral. You have, in short, a sense of acoustic space.

All this rests on the fact that your two ears—one aimed toward the left and the other toward the right—never hear the same thing in the same way. Take a pistol shot, for example. The ear aimed toward the sound source hears the shot more loudly than the other. Also, it hears the shot sooner, for the sound waves take time to travel around the head to reach the other ear. From these tiny differences in loudness and arrival time of the sound at the two ears, the brain instinctively computes direction and distance and creates a mental picture of the whole acoustic environment.

STEREO REPRODUCTION

The purpose of stereo sound equipment is to retain these space factors in reproduced music. Stereo now endows sound systems with the same two-channel principle that nature long ago designed into the head. The two separate channels of stereo equipment correspond to two separate ears. Between them they convey those small but vital differences between what is normally heard with the left and right ears (Fig. 2-1).

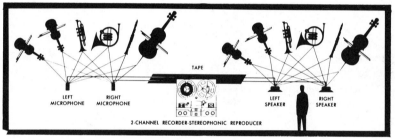

Courtesy Ampex Corp.

Fig. 2-1. Symbolic representation of stereo reproduction.

What does this mean in terms of musical experience? Suppose you are at a concert. Even with your eyes closed you sense the position of the various instruments: violins left front, woodwinds at the center, brasses at left rear, and so forth (Fig. 2-2). Many composers write their scores with these space factors in mind, and the weaving of the sound between the various sections of the orchestra—left and right, front and back—contributes to the meaning of the music. More important, stereo captures the whole acoustic aura of the original performance and transplants it into your living room, be it the spaciousness of a concert hall or the intimacy of a small cabaret. If you take a stereo recording made in a large concert hall and play it in an apartment-size room, you get the astonishing feeling that the walls are being pushed out, that the concert hall space "grows" within your living room like the genie from the bottle. The sound doesn't seem to squirt out of the speakers; instead it seems to flow broadly from the entire area between and beyond the speakers. In short, stereo delivers the music in its natural acoustic environment.

Compare this to the performance of ordinary single-channel monophonic radios or phonographs. At their best these conventional reproducers may give you the full range of musical

Courtesy Danish National Orchestra

Fig. 2-2. Stereo attempts to reproduce the sound of the concert hall.

sound, and they may do so without distortion and with good fidelity. Even so, they can render music only as a one-eared person hears it. The element of acoustic space is missing; the music is two-dimensional.

Granted, stereo sound is not a simple concept. It's easy enough to distinguish a three-dimensional object such as a bottle from a two-dimensional picture of a bottle. But to form an idea of something intangible such as sound in three dimensions is far more difficult. It is therefore not surprising that a great deal of confusion and downright misinformation persists about stereo.

Many people believe that you can get stereo simply by hooking up an extra speaker to a single-channel phonograph. Obviously this does not result in stereo, because the two speakers receive identical signals. The vital difference between the two channels (the difference between what the left and right ears normally hear) would not be reproduced. You cannot get two-eared sound from a single channel, no matter how many speakers you use.

Confusion is often expressed in questions like, "Is stereo better than hi-fi?" This reflects the notion that stereo is something to replace high fidelity. Nothing could be further amiss, for stereo has nothing to do with the basic quality of the sound you hear. Stereo merely signifies two-channel reproduction, which may be anywhere from superb to downright miserable.

For instance, you can buy a low-cost stereo phonograph. The music may spread out between the two speakers, but it lacks the quality obtained with a true high-fidelity system. On most inexpensive phonographs tonal range is clipped severely. There is no bass to lend the music a sense of fullness and warmth, and the treble is distorted. That, too, can be stereo.

Remember that stereo in itself cannot make up for basic shortcomings in the playback equipment. Stereo or no, fidelity remains the first requirement of good sound.

Three simple facts summarize the relation of high fidelity to stereo:

1. Stereo is no substitute for tone quality.
2. To be musically satisfying, stereo must meet high fidelity standards.
3. Stereo is a "plus" factor—something added to high fidelity, but fidelity remains first and foremost in importance.

Unfortunately, some of the stereo equipment offered for sale comes nowhere near high fidelity, even by the most lenient standards. This raises the questions "What is high fidelity?" and "How can I be sure to get the real thing and not a poor imitation?" It is precisely these questions that will be answered in later chapters dealing with the "hardware" of high fidelity—the equipment you need to make stereo live up to the full potential of its promise.

CHAPTER 3

Four-Channel Sound

A more recent approach to sound reproduction is four-channel sound, also known as "quadraphonic" sound, sometimes called "quad" for short. The question naturally arises: If stereo provides accurate sound reproduction, why do we need quad? Or, putting it another way, what does quad do that stereo doesn't?

The basic idea of the four-channel approach to sound reproduction is to enhance the "ambience" of sound. The term *ambience* does not refer to the sound itself. Rather it describes the acoustic surroundings in which the music is played. It is the total pattern of the sound reflections and reverberations in the concert hall or the studio. Different concert halls have their own distinctive ambience. Sometimes the producer of a recording deliberately creates certain ambience effects through microphone placement, echo chambers, or other electronic devices to enhance the impact of the music. Four-channel reproduction greatly increases the available variety of such effects.

One question that keeps coming up in regard to quad is this: Why do two-eared people need four-channel sound? The original theory behind stereo ran something like this: Each speaker corresponds to one of your ears—your two "channels" of aural perception. Consequently, two channels should be enough to reproduce all dimensions of sonic experience.

The hitch is that in real life sounds hit us from all around—not just from the front. If you are sitting at a concert, a large percentage of the total sound energy reaches you from the back and from the sides via reflections bouncing off the rear and side walls of the hall. Granted, conventional stereo catches some of this "reverb." But when you play it back, it all emerges from the speakers up front. The idea of quad is to distribute the reverb between front and rear exactly the way it occurs in a real listening situation.

This "total ambience" is produced in quad by putting two speakers in back of the listener, in addition to the two front speakers. In this way, the listener receives sonic information from virtually all directions. A complete "sound-field" is created.

COST FACTOR

Four-channel sound naturally requires four speakers and four separate amplifier channels to feed separate signals to each speaker. Therefore, the need for the extra two channels increases the cost of such systems, and the price tag for quad runs at least 50% higher than for a stereo system of roughly comparable quality. For many buyers, the primary question is: Is four-channel sound worth the extra cost?

There can be no standard answer. Everyone must decide this for himself, depending on his personal impressions. Listen to a good demonstration of quad sound, compare it with standard stereo, and then ask yourself whether the quad sound is really much more satisfying to you. Most people agree that quad definitely adds to the sonic experience. But many feel that the difference between quad and stereo is not really great enough to justify the added cost or the greater complexity of the equipment. Others are convinced that anything that adds to the impact of the music is inherently worthwhile.

To discover the difference for yourself, ask your audio dealer to demonstrate a four-channel system. Then, as you listen, switch to two-sound stereo. Every four-channel system has provisions for such switching, so you can alternate between quad and stereo as the music plays. (Of course, when you are listening to just two channels, the rear speakers should be switched off.) On the basis of such direct comparisons, you can form your personal judgment whether to "go quad" or to stick with standard stereo.

In addition to creating an accurate duplicate of the original sonic ambience of a concert, there are many other possibilities inherent in quad. Recording engineers have used the four-channel medium for bold experiments to create an entirely new kind of listening experience. On certain symphonic recordings, they placed the orchestra and the microphones in such a way that the listener seems to be located right in the middle of the orchestra, with sound impinging from all directions. With jazz combos and rock groups, recording engineers have created the impression that the musicians are moving about the listener or rushing toward him from all sides. Of course, you are not likely to hear music this way in real life—so this kind of quad sound can hardly be called realistic. But the effect can be tremendously stimulating and exciting.

QUAD PROGRAM SOURCES

Since quad requires four separate sound signals, the question is how these signals can be obtained. The simplest way to get four-channel sound is to record four parallel tracks on tape. A number of companies are now offering reel tape recorders on which four tracks can be recorded or played simultaneously. But since two extra tracks must be put on the tape, twice as much tape is used as for standard stereo. This increases cost and limits the market for open-reel tape with music prerecorded on four channels.

The simple principle of four parallel tracks has also been employed in tape cartridges. Such cartridges have been widely used in car stereo systems, and because they do have four tracks going in each direction, they were easily adaptable for home-based quad systems.

However, phonograph records are by far the most popular of all sound media, and the audio industry has devised some ingenious methods to get quad sounds from records. This meant squeezing four separate sound channels into a single record groove.

Two different principles have been employed to make four-channel records. One is called "matrixing" and forms the basis of the so-called "SQ" records manufactured by CBS and a number of other companies under their license. The other principle is called CD-4 and has been adopted by RCA and a group of companies licensed by them.

In the matrix technique, front and back channels on each side are blended into a single signal. That way, the four channels are squeezed down into two for recording. Then, in playback, a decoder in the four-channel amplifier reverses the process. It separates the blended channels again, feeding front and rear signals on each side to their respective amplifier circuits and speakers. The logic of this procedure is rather complicated, but one engineer offers a good analogy: "It's like mixing eggs and oil for mayonnaise. Then, in playback, we separate the oil and eggs again."

In the CD-4 system front and rear channels are not blended for recording but are separately engraved in the record groove. This has the advantage of providing slightly greater channel separation. However, it takes a special phono cartridge (as well as a different type of decoder) to play these disks. By contrast, the SQ matrix records can be played on any stereo record player. If the player is hooked up to a four-channel amplifier with an SQ decoder, the full quad effect is obtained. Incidentally, any quad record can be played on standard stereo equipment, but it won't sound like quad. It will simply sound like ordinary stereo. Conversely, any two-channel stereo record can be played on quad equipment. In short, the two systems are "compatible" and your stereo records will not become obsolete when you acquire quad equipment.

QUAD EQUIPMENT

To play four-channel music, whether from records, tapes, or fm quad broadcasts, you naturally need four separate speakers.

Courtesy Technics by Panasonic

Fig. 3-1. A high-powered four-channel receiver.

You also need a four-channel amplifier or receiver to drive the speakers. These amplifiers or receivers usually have the necessary decoder circuits built in for sorting out the separate channels from records or broadcasts. The high-powered receiver in Fig. 3-1 can be used to play back all types of quad programs (tapes, records, and broadcasts). Note the "outboard joystick" at the left. This control is used to adjust the balance between the four speakers.

If you play your old stereo records on such four-channel equipment, you get a bonus. Running the stereo signal from your old records through the four-channel decoder produces a pseudo-quad effect. The result is not as impressive as with records specifically made for four-channel reproduction, but a certain illusion of extra ambience is obtained. The technical reasons for this are quite complex, having to do with the phase relationships of the two channels. But getting even a trace of the four-channel effect from two-channel stereo records makes for more exciting listening. It is one of the few instances where, literally, you get something for nothing.

As far as the basic fidelity of quad is concerned, the same cautions apply here as in stereo. Quad as such is no substitute for quality. Adequate power, frequency response, and freedom from distortion still remain the chief requirements for good sound. These should never be sacrificed merely for the sake of having four channels instead of two. Or, to put it another way, a high-quality stereo system sounds better than a mediocre quad system. If you have a limited budget that won't stretch to pay for high-quality quad, you would be better off spending the available amount on a really good stereo system. After all, two good sound sources will give you more pleasure than four bad ones. Besides, you can always update your stereo system for quad later on (Fig. 3-2).

Courtesy Sansui Electronics Corp.

Fig. 3-2. A four-channel converter/amplifier for addition to a stereo system.

CHAPTER 4

Consoles or Components?

Sound systems come in two basic types: "one-lump" or "piecemeal." The one-lump variety is the *console* commonly seen in the show windows of places where radios and phonographs are sold. Their distinguishing mark is that they are fully self-contained (Fig. 4-1). To find the piecemeal type of sound system you usually have to seek out a special high-fidelity dealer, who will show you an array of separate *components*—turntables, tape players, tuners, amplifiers, and speakers—which together make up a high-fidelity system (Fig. 4-2). The question is: which should you buy, console or components?

There are two schools of thought on this subject. Many audio fans do not even seriously consider consoles. Yet many of the full-size home music systems sold are of the console type. Before you make your own choice, it would be well to consider the relative merits of the two types of systems.

CONSOLES

Consoles have one indisputable attraction: you plug them in and they play. All of the interconnecting wiring for the system has been factory installed in the console. The entire

Fig. 4-1. A fully self-contained component stereo system.

system is contained in one package; for this reason, a console is sometimes the better choice. Many consoles contain space for record storage in addition to the reproduction equipment.

Consoles are contained in furniture-styled cabinets, and many people prefer them for this reason. A wide variety of cabinet styles is available so that one can be chosen to harmonize with almost any room. In summary, shopping convenience, easy installation, and furniture styling account chiefly for the widespread appeal of console radio-phonographs. In terms of sound quality, consoles can be as good as component systems. A first-rate console very likely sounds superior to a second-rate component system. But first-rate consoles are usually expensive because the cost of the cabinet must be added to the cost of the reproducing equipment.

If your preference is for a console, you can find a number of fine-sounding models among the top lines of reputable manufacturers. But keep in mind that it requires careful shopping and attentive listening comparisons to find audio equipment that suits your particular requirements.

Fig. 4-2. A typical component system placed on open shelves.

COMPONENT SYSTEMS

One of the chief advantages of component systems is flexibility of installation. A component system need not take up any floor space at all. The components, including the speakers, can be placed on shelves, room dividers, etc. As for appearance, current component styling harmonizes with contemporary settings; so if you have modern decor, you can usually leave your components sitting out on the shelf and save the cost of cabinets altogether. Or you may mount the components in furniture you already have—cupboards, chests, etc. That, too, saves both space and money and assures you that your sound system harmonizes with the decorating scheme of your home.

Components can be combined at will in an add-on pattern to let you build up your system gradually, thus reducing the initial cost. For instance, you may start with a stereo record-playing setup and later add a tuner to receive fm broadcasts. Finally, perhaps, you may augment this with a tape player or recorder. Components permit you to add these features one at a time.

COMPARISON OF SYSTEMS

An important question to consider is how the sound system will fit into your home. An argument in favor of consoles is that they are self-contained. However, to set the speakers far enough apart for effective stereo, a console should be at least 6 feet long and preferably even longer. This makes it a rather bulky item, especially for rooms of moderate size.

In their physical layout, components allow greater latitude. For instance, you can locate the speakers for optimum stereo effect right where you sit in your favorite chair. And, if you like, all the controls can be placed next to your chair while the sound comes from the speakers all the way across the room. That way you won't have to jump up every time you want to flip a record, tune to another station, or change the volume. A few consoles have remote control facilities, but they involve additional cost.

If consoles and components represent generally different design philosophies, it is nevertheless possible to make the best of both worlds in so-called component consoles. These are made by some of the component manufacturers who, optionally, place their components in prepackaged furniture units containing the entire system. At the sacrifice of physical flexibility of installation and of choosing each component separately, these component consoles offer an opportunity to buy components with clearly specified capabilities as a complete single-unit piece of furniture.

Ordinary consoles differ from component systems in an important technical sense. A component system is like a doctor's prescription—everything that goes into it is specified clearly. For each individual component, you get a specification sheet that states its exact capabilities. The purpose of this book is to give you some basic guidelines for interpreting these specifications so you can pick knowingly the components best suited to your particular needs.

The very fact that the component manufacturer, by giving you these specifications, is willing to lay his cards on the table before you is an assurance of value. You are dealing with known quantities. In most cases, published specifications take the guesswork and gamble out of buying. The situation is different when you buy an ordinary phonograph or console. Only rarely does the manufacturer tell you about the capabilities of his set in precise technical terms. He may tell you how many speakers it has, or how many watts the amplifier puts out. Seldom does he specify the exact frequency response along with the percentage of distortion at various output levels.

You might think of it this way: unless full specifications are given, you are taking potluck with whatever the console manufacturer has put inside the cabinet. But when you buy components, it's like ordering *a la carte*. You choose each item to meet your particular demands in performance, features, looks, and price.

COMPACT SYSTEMS

Within the last few years, a new type of stereo equipment has been introduced which may be regarded as something halfway between standard component systems and ready-made consoles. Commonly called "compact systems," these systems are hardly bigger than ordinary table phonographs, but their clear, full-bodied sound nearly rivals the fidelity formerly found only in full-sized component systems.

The basic idea behind the new compacts is a three-piece design as shown in Fig. 4-3. One unit contains all the "works" —record changer, amplifier, radio tuner, and controls. The other two units are left and right stereo speakers. Since they are detached from the main unit, you can put them wherever they sound best. Connecting the speakers to the main unit is as simple as plugging in a lamp.

In effect, the compact systems combine the flexibility of components with the simplicity of a console. And despite their

Courtesy Harman-Kardon, Inc.

Fig. 4-3. Self-contained stereo system including a-m and fm tuner, record changer, and power amplifier.

small size some of the better compact systems rival the performance quality of regular components.

CHAPTER 5

Anatomy of a Sound System

A sound system is not a single piece of machinery. Its basic anatomy always consists of separate components, even if you buy them lumped together in a console. Functionally, these components can be compared to the organs of the human body. Each is assigned a specific job, and jointly—in carefully planned division of labor—they perform the many different tasks involved in sound reproduction. First, each of the components will be defined, and just what it does will be briefly explained. Later the pertinent design factors of each component will be discussed in greater detail.

TURNTABLE

Just going in circles seems easy enough, but spinning a record is an exacting job. Correct speed must be maintained within close tolerances; otherwise, a wavy kind of sound and unsteady pitch known as *wow* is produced. Moreover, the turntable must rotate without any trace of chugging, or else the music suffers from a vibrato effect known as *flutter*. Equally important, the turntable must be free of vibration, for any vibration would be amplified along with the music and reach the ears as a low growl, similar to the sound of a distant railroad train. This is called *rumble* and afflicts the majority of turntables that fall short of high-fidelity standards. To avoid these various pitfalls, high-fidelity turntables must be precision-machined and use extremely steady, smooth-running drive motors. Fig. 5-1 shows a typical turntable and its tone arm.

TONE ARM

The job of the tone arm is to guide the cartridge across the record. To achieve high fidelity, the tone arm must move freely with very low frictional drag; it must have stable balance so as to accurately maintain proper pressure on the stylus; and

Courtesy Acoustic Research, Inc.

Fig. 5-1. High-fidelity turntable and tone arm.

it must have no audible resonance. Moreover, the dimensions and angles of the arm must be carefully calculated to keep the cartridge as closely as possible in a line tangent to the record groove, regardless of which part of the record is playing. An arm failing to meet these requirements causes distortion, groove jumping, and record wear.

RECORD CHANGER

Instead of using a separate turntable and tone arm, some people prefer automatic record changers. It should be pointed out, however, that changers do not play both sides of the same record in succession. So, to play a symphony recorded on two sides of the same record, it is still necessary to get up and turn over the record.

Some record changers do not meet high-fidelity standards. However, high-quality changers, usually labeled as "automatic turntables", are available for those who wish to combine high-fidelity performance with automatic record-changing functions.

CARTRIDGE

The cartridge (Fig. 5-2), sometimes called the *pickup*, is mounted at the tip of the tone arm. Through its stylus—which corresponds to the needle in old phonographs—it traces the sound vibrations engraved in the record groove and translates

Fig. 5-2. A typical cartridge.

Courtesy Shure Brothers, Inc.

these vibrations into electric signals. The quality of the cartridge depends on its ability to follow the record groove with great precision and to reproduce all frequencies in their true proportions.

AMPLIFIER AND PREAMPLIFIER

The amplifier receives from the cartridge very weak electric signals representing the music. It then enlarges, or amplifies, these signals to make them powerful enough to drive the speaker and fill the room with sound. This "enlargement" of the signal must be a precise replica of the original. Any deviation from the original signal waveform provided by the cartridge would be distortion, falsifying the true sound of music. Low distortion, among several other factors, is therefore a chief requirement of a high-fidelity amplifier.

An amplifier consists of two sections: the *preamplifier* and the *power amplifier*. The preamplifier (called "preamp" for short) contains the various controls that let you choose your program source (records, radio, tape, etc.), adjust volume, stereo balance, treble and bass, and perform various other control functions. The power amplifier takes the signal from the preamplifier and builds it up to the strength (output power) necessary to drive the speakers. This output power is rated in watts.

In amplifiers of moderate power, the preamplifier and power amplifier are usually combined in a single unit (Fig. 5-3). In high-power equipment the power amplifier and preamplifier may be separate pieces of equipment.

Courtesy Sansui Electronics Corp.

Fig. 5-3. A combination amplifier and preamplifier.

SPEAKER

The speaker (Fig. 5-4) is the voice of your sound system—the only component that actually generates sound. Taking the amplified signal from the amplifier, the speaker converts it back from electricity into audible music. More than any other single component, it determines the quality of the sound you hear.

Since it is difficult for a single speaker unit to span the whole musical range, special speakers called *woofers* are generally used for the low notes. At the high end of the scale, high-frequency speakers known as *tweeters* take over. Today most speakers are sold as complete speaker systems, combining woofers and tweeters in a matching enclosure. Hence, you need not select woofers and tweeters separately.

Like most other components, speakers should have a flat frequency response; i.e., they should reproduce all tones equally

Fig. 5-4. A typical high-fidelity speaker.

Courtesy Electro-Voice, Inc.

well without unnatural emphasis or suppression of any of the frequencies that make up the music. This is harder to achieve in speakers than in other components and, therefore, is a very critical requirement met only by high-quality speaker systems.

TUNER

As its name implies, the tuner (Fig. 5-5) is a device that lets you tune in radio stations and play their programs through your high-fidelity system, which naturally reproduces the broadcasts with a sonic quality far superior to that of ordinary radios. Because a-m reception is often limited in fidelity and subject to static, most audio fans rely on fm for their radio

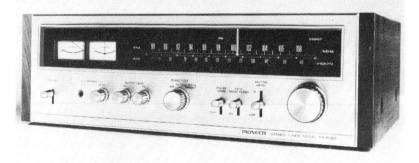

Courtesy U. S. Pioneer Electronics Corp.

Fig. 5-5. A separate am/fm stereo tuner.

entertainment. A good fm tuner provides reception of a quality that satisfies the most stringent high-fidelity requirements.

Most fm stations broadcast in stereo by a process called *multiplex,* which permits a single fm transmitter to send out both stereo channels. Some of these stations also broadcast four-channel programs by the same method. On ordinary stereo equipment, these four-channel programs are received just like a standard stereo broadcast. However, with a four-channel receiver (which includes a special four-channel decoding circuit) all four channels are reproduced separately.

RECEIVER

For simplicity and compactness virtually all manufacturers offer as an alternate choice a single unit that combines the functions of tuner, preamplifier, and power amplifier as illustrated in Fig. 5-6. These are called receivers or tuner-amplifiers. Only record-playing equipment (turntable, tone arm, and cartridge) and a speaker or speakers need be added to complete the sound system. A tape player may be used in addition to or instead of the phonograph turntable.

The key to successful receiver design is the use of transistors, which in recent years have completely revolutionized the basic concepts of audio engineering. The compactness of transistor circuits has enabled engineers to cram all the electronics of a stereo system into a single, handy package without sacrifice of quality.

Courtesy Electro-Voice, Inc.

Fig. 5-6. An elegant stereo fm tuner and amplifier with 25 watts per channel.

These new design concepts account for the popularity of stereo receivers, which are now in greater demand than separate amplifiers and tuners. Ease of installation, compactness, convenience, and economy are the main reasons for the present shift to stereo receivers as the dominant type of component.

TAPE RECORDER

The tape recorder (Fig. 5-7) is not usually considered a necessary item in a sound system. Rather it is an extra to add such functions as recording off the air, copying disc recordings, or making your own live recordings via microphone.

If you do not plan to take your tape recorder to record in various places, but intend to leave it at home as a permanent part of your sound system, it need not contain any speakers of its own, for you would always play your tapes through your regular speakers. For such applications you can get tape machines minus built-in speakers. These are known as *tape decks*. Tape decks come in three different types—open-reel, cassettes, and cartridges. Their respective advantages and drawbacks will be discussed later.

Courtesy TEAC Corp. of America

Fig. 5-7. A typical high-quality stereo tape recorder.

CHAPTER 6

Quality Factors in Components

The basic tasks of each component were outlined in Chapter 5. The performance standards each component must meet in order to qualify for high-fidelity sound reproduction will now be discussed. Some of the factors that distinguish components of high quality have already been mentioned. Now these requirements will be discussed in more detail.

From a technical point of view, components may be divided into three classes: the purely electronic (amplifier and tuner); the electromechanical (components combining both electrical and mechanical functions), such as speakers and cartridges; and the purely mechanical, such as turntables and tone arms.

AMPLIFIERS

The heart of any high-fidelity system is the amplifier. Some of the requirements for a good amplifier will now be discussed.

Power Rating

The most widely advertised specification of an amplifier is its *power rating*, or *output power*. The two terms mean the same thing and are always expressed as a certain number of watts. This tells you how much power the amplifier is capable of delivering to the speakers. A very common misconception is the idea that the number of watts of output power tells how loud an amplifier can play. This is like saying that the amount of horsepower determines how fast a car can go—it is only partly true.

For instance, if you had a 60-watt amplifier and a 30-watt amplifier both playing at top volume, the 60-watt amplifier would sound only a little louder than the 30-watt amplifier—not nearly twice as loud. The reason for this lies in a rather complicated logarithmic relation of physical energy in sound to perceived loudness—a factor built by nature into human hearing. For practical purposes all you need remember is that doubling output power doesn't mean doubling the loudness.

But there is some justification in comparing output power in amplifiers to horsepower in cars. In both cases it's a matter of sufficient power reserve for critical moments. You do not always run your car with the gas pedal all the way to the floor, extracting every bit of available power from the engine. Similarly, the amplifier only rarely operates at full output. But there are moments in music—just as there are moments on the road—when ample reserve of power helps you over the hurdles. Such musical "hurdles" are orchestral climaxes—crashing fortissimos, loud chords forcefully struck on the piano, or the deep rolling notes of the low bass. When you consider the sheer physical effort that musicians put into the playing of such passages, it is easy to see that such sounds represent a tremendous concentration of physical energy. It is in such passages that an extra margin of power is needed for clear reproduction.

Suppose you have a rather small amplifier—about 10 watts per channel. Playing music at room-filling volume would not necessarily overtax such an amplifier, but it would leave relatively little power reserve if the room were fairly large. Chances are that the music would sound very agreeable for most of the time. But suppose the score calls for a sudden triple forte—for example, a fanfare in brass. It would sound loud all right, but because of the limited power reserve available for this extra surge, the amplifier would momentarily veer into distortion, and at the climax the sound would become harsh. This may last only for an instant. As soon as the music subsides to normal levels, the amplifier returns to its normal behavior. But the human brain subconsciously retains memories of these momentary distortions, and you begin to feel vaguely dissatisfied. This is known as "listener fatigue"—a psychological reaction to distorted sound.

By contrast, an amplifier with ample power reserve takes all climaxes in its stride without budging into distortion. And again you react subconsciously to the quality of the sound. The uniform, unbroken clarity of the music increases your sense of pleasure in listening. You feel relaxed and more receptive to the music—no psychological "listener fatigue" occurs.

How much output power suffices to achieve this happy condition? Unfortunately, no standard answer fits all cases. Your exact power requirement depends on the size of your listening room, the kind of speakers you use (some speakers need

Courtesy Sansui Electronics Corp.

Fig. 6-1. An integrated stereo amplifier with four-channel options.

more watts than others to produce a given loudness), and the kind of music you like and how loud you like it. Even your furnishings must be taken into account. A later chapter will formulate some rules on calculating all these factors so that you can arrive at a definite wattage figure to fit your particular needs. But for rough guidance, 20 to 25 watts per channel is ample for most situations. Greater power usually is needed only in very large rooms and for very heavily scored orchestral music critically listened to at loud volume. At the low end of the power scale, 15 watts per channel might be considered a satisfactory minimum for apartment-size rooms. Fig. 6-1 shows a typical high-fidelity amplifier.

There are various ways in which wattage is expressed in amplifier applications. What is labeled as a 50-watt amplifier is usually an amplifier that delivers 25 watts per channel. True, it delivers a total of 50 watts, but it must be remembered that in terms of listening the two channels don't really add up that way, for each channel operates independently and must by itself furnish sufficient power to drive its own speaker.

So it's the 25-watt figure—the watts-per-channel rating—that really counts when figuring your power requirements.

Unfortunately, no single standard method for measuring amplifier power has been adopted, and different methods yield different wattage figures. For example, an amplifier rated at 20 watts per channel by strict measurements may appear as an 80-watt amplifier by more lenient methods. Naturally, some manufacturers (especially those of poor equipment) take advantage of such discrepancies in their advertising claims to make their product seem better than it actually is. Some unscrupulous advertising practices border on fraud in this respect. Reputable manufacturers of high-quality audio equipment have therefore agreed to state output power by one of two reliable methods. One is called the *IHF standard* (set by the Institute of High Fidelity). The other is the even more exacting *rms method* (rms stands for "root mean square"—a strict engineering method for measuring audio power. Whenever a wattage figure is followed by either of these two abbreviations (IHF or rms) it indicates that trustworthy measurement methods have been employed.

Distortion

Have you ever looked at yourself in a funny-house mirror? It perfectly illustrates the principle of distortion—a replica still recognizable but untrue to the original (Fig. 6-2). The same thing happens to music in bad sound equipment—the true waveform of the musical signal gets twisted out of shape. Music is notably prone to such distortion because its waveforms are extremely complex. To preserve their meticulously detailed form is quite a challenge to the science of electronic design.

At its worst, distortion can make a violin screech like a trolley car rounding a curve. Far more often distortion is quite subtle and barely perceptible. But its effect, like that of the Chinese water torture, is cumulative. After an hour or so of

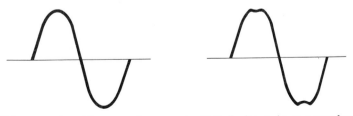

(A) Representation of input waveform.   (B) Badly distorted output waveform.

Fig. 6-2. Pictorial illustration of distortion.

listening you again feel the uneasy symptoms of listener fatigue. You may not be aware of distortion as such, but you somehow get the urge to turn the music off. But if your equipment is free of distortion, you will find yourself listening for hours on end with undiminished pleasure.

Fortunately, distortion is easily tracked down in the control laboratories of high-fidelity manufacturers. Distortion can be accurately measured and stated as a percentage indicating the proportion of distorted sound in the total output of the amplifier. Such distortion measurements are a most important index of amplifier quality.

Distortion exists in two basic forms. One is *harmonic distortion*, which despite its name can sound quite unharmonious. Harmonics, as you may remember from the chapter on the nature of sound, are the overtones of a given fundamental frequency played by an instrument. Harmonic distortion occurs when the amplifier (or any other component) adds some overtones of its own that are not contained in the original music. Such gratuitous contributions falsify tone color because tone color is determined by the harmonics (overtone structure) of a given sound.

The other form of distortion is *intermodulation distortion*, called IM for short. It is caused when two or more different notes pass through the amplifier at the same time, particularly if the notes are far apart in pitch (e.g., a bass fiddle and a flute playing simultaneously). The dissimilar frequencies tend to interact within the amplifier—colliding with each other, so to speak. From this collision spring fragments of new, discordant frequencies that are not part of the original sound. The sum of such "intermodulation products" adds harshness and fuzziness to the reproduced sound.

Luckily, modern amplifier circuits incorporate a feature known as *inverse feedback* which acts as a kind of self-checking device on the amplifier and, in the best designs, keeps both kinds of distortion at a virtually unnoticeable level. In top-rated amplifiers both harmonic and IM distortion are kept well under 1% at full output.

Power output and distortion, the two specifications so far discussed, are always related. It is customary to make distortion measurements at the rated maximum power of the amplifier. Distrust any distortion figures that do not specify the power level at which they were taken. Jointly, distortion figures and output power give a pretty good picture of an amplifier's capabilities. That is why these two quantities are often presented as a graph in which one is plotted against the other (Fig. 6-3).

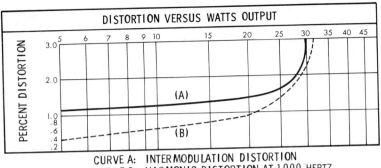

Fig. 6-3. Graph showing distortion versus power output for a hypothetical amplifier.

Frequency Response

Of all high-fidelity specifications, frequency response is perhaps the one most freely bandied about and also the most widely misunderstood. Much emphasis is placed on wide-range response, and specifications may proudly claim response to the far ends of the audible spectrum—from about 20 to 20,000 hertz (cycles per second). But such a statement is almost meaningless. It merely tells the top and bottom notes the amplifier (or any other component) is capable of reproducing. What is more important is the uniformity of response between those extremes. The amplifier (or, for that matter, any of the other components) must reproduce all notes in precisely the same proportion as they are heard in the concert hall or studio. Low, middle, and high notes must all keep their true relation to one another. The equipment must not emphasize some notes more than others or swallow certain notes—every note must be given precisely its due value. This is what is meant by the often heard term *flat frequency response*.

The term "flat" frequency response, or flat response for short, originates in the practice of showing frequency response in the form of a graph plotting the intensity of each reproduced test tone against its frequency (Fig. 6-4). If the test tones are all of the same intensity to begin with, the equipment should reproduce them all with the same intensity. When such an ideal response is drawn on graph paper, you come out with a "flat" straight line.

Perfectly flat response remains a pipe dream of audio engineers, but the best of modern amplifiers come remarkably close to it. To tell just how close, you should always look for a "plus-minus" figure (prefixed by the sign $\pm$) following the statement of overall frequency range. For instance, frequency response may be specified as 20 to 20,000 hertz ±1 dB. This means that at no point in the whole range from 20 to 20,000 hertz does response deviate by more than 1 dB from true "flat-

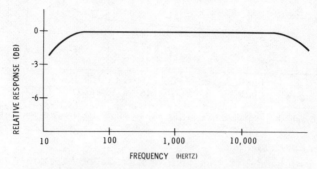

Fig. 6-4. Graphical representation of flat frequency response.

ness." (The dB stands for "decibel" and is the common measure of sound intensity or loudness.) A frequency response with a maximum deviation of ±1 dB may be considered "flat" for all practical purposes. In the best of modern amplifiers, the maximum deviation from flat response is as low as ±0.5 dB over the whole audible range.

Some manufacturers give two separate specifications for frequency response—one at the nominal level of 1 watt, the other at maximum rated output. The reason for this is that extreme highs and lows are more difficult to reproduce at full power than the middle of the scale. An amplifier able to cover the entire range with flat response even at maximum power usually has a margin in clarity and depth of sound at any volume level.

Transient Response

The term transient refers to sounds that appear with sudden impact and cease just as suddenly. Drum beats and other percussive sounds, crashing piano chords, or plucked strings are cases in point. But not all transients are so obvious; in fact, such subtle sonic details as staccato are also transient in their physical character.

Such sounds present problems to the amplifier, as well as to other components, because of their abruptness. If this abruptness is lost in reproduction because the equipment does not respond quickly enough to the sudden onset of sound, the music loses a good deal of clarity and definition. With poor transient response, the entire orchestra, instead of seeming excitingly alive, sounds a little blurred and soggy. In sum, transient response describes the ability of the sound equipment to handle

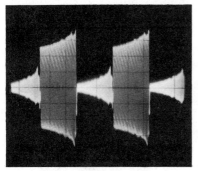
(A) Poor transient response.

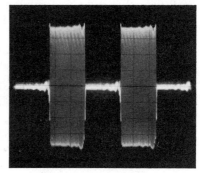
(B) Good transient response.

Fig. 6-5. Transient response in speakers as measured by tone bust.

these brief bursts of power that constitute the transient waveform of the sound.

Direct measures of transient response are rarely stated in the specifications because no absolutely objective method of measurement exists. Some manufacturers show so-called square-wave patterns in their qualifications. These are photographs showing how a rectangular square wave (its vertical side representing the sharpest of all possible transient waveforms) is reproduced by the amplifier. The photographs are pictures of an oscilloscope screen on which the amplifier output is displayed electronically in the form of a graph as shown in Fig. 6-5. Such photographs give a fair indication of transient response, but they are subject to subjective interpretation and provide reliable information only to trained persons experienced in relating the visual waveform to audible results.

Fortunately, there is a simpler way of guessing at, if not measuring, the transient response of an amplifier. As a rough rule, the wider and flatter the frequency response and the higher the power reserve, the better the transient response will be. That is why the frequency response of the best amplifiers extends far beyond the limits of human hearing. Even though you can't hear anything above approximately 16,000 hertz, flat amplifier response to 20,000 hertz and beyond pays off in added clarity of sound. The transients are more truthfully reproduced. Some amplifier designers also extend the low-frequency limit below the range of human hearing, hoping to attain sharper definition and greater transparency of sound. Thus, certain top-line amplifiers have virtually flat frequency response extending all the way from 5 to 100,000 hertz at the nominal 1-watt level. But flat response from 30 to 20,000 hertz may be considered an acceptable minimum standard for high fidelity.

As we have mentioned, transients represent a sudden concentration of power. Therefore, sufficient margin in amplifier power is also a help in handling these vital but elusive details of musical sound.

Hum and Noise

Perhaps the most underrated pleasure of high fidelity is to hear the music reproduced against a background of relative silence. Whenever ordinary radio-phonographs are turned on, the first thing you hear is the hum of the electric house current. Most of us are so used to this that we take it almost for granted to hear reproduced music obscured by a certain amount of unrelated noise. This makes the realism of music emerging from the speaker without such noise all the more startling.

The power-supply sections of high-fidelity amplifiers and tuners are carefully designed to keep hum at a minimum by filtering the supply voltage through large (and expensive) filter capacitors. Moreover, hum-sensitive parts are carefully shielded.

Hum and noise ratings are stated in decibels (dB). For instance, a specification saying "hum and noise: 50 dB below full output" means that the hum and other inherent noises of the amplifier are 50 dB softer than the musical signal reproduced at full output power. This is quite unobtrusive, but some top-rated amplifiers attain even better signal-to-noise ratios, up to -80 and -90 dB, which is completely inaudible. As a rule, the hum and noise figure for the phono input (i.e., for record reproduction) is not quite as favorable as for other program sources (tuner or tape recorder) because the greater amplification necessary for record reproduction entails a somewhat higher noise level. However, a ratio of -50 dB for the phono input may be considered good, and -60 dB excellent.

TUNERS

In many areas, with an fm tuner hooked into your sound system you can have an inexhaustible, free supply of good sound. Audio fans have long been aware that fm is the only broadcasting method capable of meeting the technical standards of high fidelity. Ordinary a-m radio usually does not yield the required frequency range, quietness of background (absence of static), and dynamic range (difference between loud and soft), but fm excels in all these respects. Moreover, a great many fm stations offer musical programming—both "light" and classical—chosen with good taste and designed to appeal especially to the more discriminating listener. Most fm stations broadcast in stereo, sending out both stereo channels from a single transmitter by means of the fm-multiplex process.

Types of Tuners

Some tuners provide fm reception only. However, most of the tuners produced today combine a-m and fm reception, as illustrated in Fig. 6-6. Of course, you can normally get high fidelity only on fm, but the a-m section lets you tune in to programs that might not be available on fm in your locality. The trend in modern equipment is to complete receivers, instead of separate tuners, amplifiers, etc. Such a receiver is pictured in Fig. 6-7.

Tuner Specifications

Sensitivity is the most widely publicized tuner specification. It describes the ability of the tuner to pull in weak and distant stations. Sensitivity is not of primary importance if you live near the fm stations you want to receive. But if you are in a fringe reception area, sufficient sensitivity can make the differ-

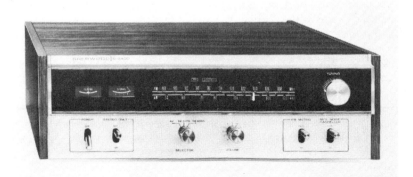

Courtesy Sherwood Electronic Lab.

Fig. 6-6. An am/fm stereo tuner with dual meters.

Courtesy Kenwood Corp.

Fig. 6-7. A compact combining tuner, amplifier, and controls. Rated at 90 watts per channel (IHF).

ence between satisfactory and marginal reception—especially in stereo, which requires greater signal-pulling ability than mono.

Sensitivity is always stated in relation to *quieting*. Quieting refers to the ability of the tuner to strip off noise from the radio signal so that nothing but a clear, undisturbed audio signal appears at the tuner output to be fed into the amplifier. Quieting is specified in decibels (dB). If the specification reads: "Sensitivity: 3 μV for 30-dB quieting," it means that an incoming signal must be 3 microvolts strong at the antenna terminals in order to quiet the noise to a level 30 dB below the music.

The Institute of High Fidelity (IHF) suggests that all sensitivity measurements should be made for 30-dB quieting (this is part of the requirement for measuring the so-called usable sensitivity). The majority of manufacturers observe the norm. Others, however, use a less stringent yardstick that requires only 20-dB quieting for whatever sensitivity figure is stated. When you compare sensitivity ratings, you should watch for the quieting figure that accompanies the statement. If it says "20-dB quieting," you know that the tuner is less sensitive than one rated for 30-dB quieting, even though the nominal sensitivity figure is the same. If the sensitivity figure is in no way explained, don't trust it. If the sensitivity rating is followed by the letter IHF, it means that the IHF standard of 30-dB quieting has been observed in the measurement.

As a general rule, a tuner with 4- to 7-μV sensitivity should do very well if you live close to the station, but beyond a distance of 25 miles or so, 3- to 5-μv sensitivity might be considered more advisable. In fringe areas, sensitivity ratings of 1.5 to 2 μV may be necessary for clear reception.

Of course, even the most sensitive tuner is handicapped unless provided with a good antenna. Near the station, a simple indoor antenna may suffice, but in more distant locations a roof antenna usually brings considerable improvement in clarity of sound and reduction of background noise. In fringe areas, a multielement yagi antenna is virtually a must.

Fm transmission is subject to the same physical laws as tv transmission. Steel buildings, mountain ranges, and other terrain features may partially block signals. In low-lying valleys, reception is usually difficult. In such cases a multielement antenna may be aided by a so-called antenna booster (a small amplifier located atop the antenna mast). If the various stations you want to receive lie in different directions from your home, an antenna rotator may be necessary to face the antenna toward the station you want.

Intermodulation and harmonic distortion in tuners should not exceed 1% and in top-quality tuners may be less than 0.5%. These distortion measurements should apply to 100% modulation, i.e., full tuner output. Audio-frequency response should be flat within ±1 dB to at least 15,000 hertz, which is the upper frequency limit of fm transmissions.

You will find among fm tuner specifications such entries as discriminator bandwidth, i-f bandwidth, number of i-f stages, and various circuit configurations. Without detailed knowledge of radio theory these factors are difficult to comprehend, and an explanation exceeds the limits of this basically nontechnical introduction. Fortunately, none of these factors are of direct importance to the high-fidelity user, because their joint effect on performance can be told in terms of the specifications we have already discussed: sensitivity, distortion, and frequency response.

However, one added tuner specification, the so-called *capture ratio*, may be important to you if you live in a location where two fm stations are picked up at the same spot on the dial. This rarely happens if you live near a city, for in any given locality the various fm stations are spread out over different parts of the dial. But if you live midway between two cities you may conceivably find yourself with two stations coming in on the same frequency, making it impossible to listen to either of them. Good capture ratio, however, enables the tuner to concentrate on the stronger of the two signals, rejecting the weaker one. Capture ratio is expressed in decibels signifying the minimum difference in signal strength that enables a tuner to hold two conflicting stations apart. Hence, the lower the figure, the better the capture ratio.

Capture ratio, by the way, should not be confused with selectivity. Selectivity refers to the ability of the tuner to keep apart stations adjacent to each other on the dial. Capture ratio refers to its ability to suppress one of two stations at the same place on the dial.

Operating Features

The most important single factor in operating your tuner, and your best guarantee of good sound, is accuracy of tuning. You get optimum sound quality only if you tune in your station at precisely the right spot. Careless tuning distorts the sound in loud passages and virtually wipes out stereo separation if you are listening to an fm-stereo station. To assure tuning accuracy and ease, some kind of visual tuning indicator should be part of any good tuner. Then you don't have to hunt for the station by turning the tuning knob back and forth. A pointer on a meter or the green beam of an indicator tube tells you exactly when your station is in perfect "focus." You can even tune to the station in complete silence with the volume all the way down. Then, after the visual tuning meter has marked the right spot, you just turn up the volume—and right from the start your station comes in clearly.

Another convenience provided by many fm tuners is *automatic frequency control* (abbreviated afc). This is a circuit that "locks" in the station and keeps it from drifting out of tune. On cheaply made fm radios you may have had the experience of a station becoming fuzzy after a while. Then you had to get up and retune the receiver. The purpose of afc is to hold the station steady once it is tuned in. Some manufacturers strive to make their tuners drift-free to such an extent that the lock-in action of the afc is not needed. Hence, the absence of afc does not necessarily imply that the tuner will drift, but it makes accurate tuning even more important.

Many tuners have additional features, such as squelch circuits to silence the rushing noise heard between stations, pilot lights to indicate whether or not a station is broadcasting in stereo, and some even have double tuning meters—one to indicate the strength of the incoming signal, the other to show the precise tuning position. All these features are for operating convenience and have no bearing on the basic capabilities of the tuner, which are determined by the main specifications discussed earlier.

SPEAKERS

In a sense, a speaker could be compared to a musical instrument since it is a reproducer of musical tones. Its purpose, after all, is to make musical sounds just like those of true instruments. The design of speakers, involving vibrating elements operating in acoustically matched enclosures, has much in common with the design of many musical instruments. The arts of electronics and of instrument building clearly overlap in this area, and for this reason purely technical judgment of quality is not applicable. Elements of personal taste and individual variation enter into consideration.

For instance, no two makes of speakers sound exactly alike—just as no two violins or two pianos are exactly identical in sound. But these differences do not necessarily mean that one speaker is better than another. Neglecting differences due to outright deficiencies, divergence in sound among different speakers of comparable quality often reflects legitimate variables of personal taste. One designer may favor a warmer, more mellow sound while another stresses brilliance and brightness. You thus have an opportunity of choosing the kind of tonal coloration that most closely meets your own preference.

Speaker design abounds in subjective factors that elude precise measurement. Hence, numerical specifications provide no conclusive indication of how a speaker really sounds, and listening tests are the only really meaningful method of evaluation.

The most important single factor is clarity. Put on a good record; make sure it is not worn, not dusty, and not itself strident in sound. Listen carefully to the speaker. Is the sound really smooth, free from fuzziness and harshness? Is it free of that edgy quality that often mistakenly passes for "hi-fi" on cheap equipment? And above all, does it keep orchestral passages unmuddled and clearly transparent?

Distrust any speaker that sounds spectacular with big, boomy bass and sharp treble. Sounding bigger than life is the telltale mark of many inferior speakers. What seems so impressive at first listening is often due to resonances that falsely emphasize bass and treble. You will soon tire of these exaggerations and long for less obtrusive, truer sound. A really superior, natural-sounding speaker nearly always seems less dramatic at first, but it has an ingratiating quality of smoothness and "sweetness" of sound. It is a good idea to go to a concert shortly before buying a speaker, to refresh your memory of the true sound of music.

Clarity in a speaker, as in an amplifier, is mainly the result of low distortion. Distortion measurements are rarely available for speakers because there are no standard methods for measuring distortion in speakers. Even if such figures are given (they sometimes are stated for first-rate speakers), they cannot be directly correlated to the subjective impression on the ear, which remains the final criterion.

Extended frequency response, so widely advertised, is actually less important than clarity and absence of distortion in

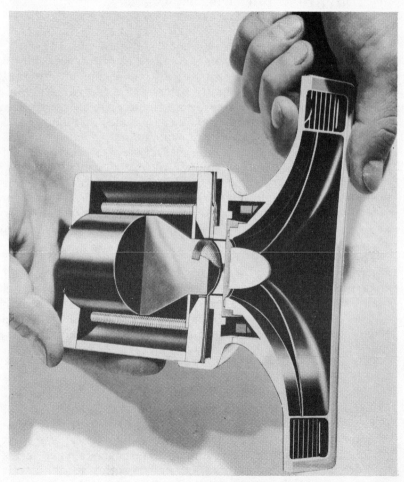

Fig. 6-8. Cross-section of a typical tweeter.

achieving natural sound reproduction. More important than extended frequency range is a natural balance between high and low frequencies.

Treble Response

Virtually all speakers employ separate tweeters (high-frequency speakers) that extend the upper range to the limit of human hearing (Fig. 6-8). Just because treble response is so far extended, it is important that the frequency response be uniform. Any sharp deviation from flat response is known as a response peak (if it emphasizes certain notes) or a response dip (if it suppresses certain notes). Peaks can be highly troublesome in speakers, imparting to the sound harshness and a feeling of constriction. It is difficult, however, to eliminate all peaks in speakers since there are natural resonances to contend with. Even the best speakers, therefore, have a rather uneven frequency-response curve, but as long as the peaks do not exceed about 5 dB, they are not considered serious if the overall effect remains smooth and pleasant to the ear. For this reason manufacturers rarely publish frequency-response curves for their speakers.

Bass Response

The very lowest reaches of the musical frequency spectrum lie below 40 hertz. True, not many notes are ever written below 40 hertz, but if a speaker is capable of reaching below 40 hertz, it has a useful margin of bass response that lends an added feeling of weight and force to ponderous bass tones. Good response in this range, however, is difficult to attain, particularly in the compact "bookshelf" speakers that are currently so popular (see Fig. 6-9). The specifications may claim response to 30 or 35 hertz, but such statements are meaningless unless the deviation from flat response is also stated for that frequency (for instance, −5 dB at 30 hertz). A speaker may, in fact, respond to very low notes, but the response may be too weak at these frequencies to be effective.

The best of compact bookshelf speakers provide bass that is full and rich despite their relatively small size. Unfortunately, many low-cost bookshelf speakers are lacking in that fullness of bass that gives orchestral sound its true balance. Again, listening comparisons are the best way to assess the bass projection of a speaker.

The common practice of playing organ music to test the bass response of a speaker is based on a fallacy. True, the organ

Fig. 6-9. Lineup of speakers in a bookshelf-type system.

bass sounds impressive, but that is just what flatters the speaker out of proportion to its true merit. A more revealing test is to play an orchestral passage heavily scored for cellos and double basses, with occasional thuds of the big bass drum. Observe whether the deep, heavy quality of these instruments is truthfully conveyed and whether the bass drum retains its sonority without causing the speaker to "break up" into distortion. Many experienced listeners consider such passages a more realistic test of speaker performance than the separate test tones contained on frequency test records.

With the extended high-frequency response common in today's speakers, adequate bass is necessary to retain a suitable tonal balance. Merely to extend the high-frequency response without having sufficient bass as a counterpoise makes the music seem thin and shallow.

Directionality

An important but all too often neglected performance factor in speakers is *directionality*. In highly directional speakers the treble frequencies are not properly dispersed; rather, they emerge as a narrow beam like the beam of light from the headlights of a car. Invariably, this creates a sense of sonic constriction, making the sound tight and lacking in natural spaciousness. One way of testing the treble dispersion of a speaker is to walk across the room and observe if the tone color of the speaker changes abruptly in different parts of the room. Compare the sound directly in front of the speaker ("on axis") with the sound you get standing about 60 degrees

Courtesy Acoustic Research, Inc.

Fig. 6-10. Dome-shaped diaphragm midrange and tweeter units designed to provide wide-angle dispersion of high frequencies.

to side. The speaker with the most even treble dispersion usually not only sounds better, but is also more easily adaptable to the acoustics of different rooms.

Dome-shaped diaphragms (Fig. 6-10) on midrange and tweeter units are designed to provide wide-angle dispersion of high frequencies. From the vibrating dome, sound radiates hemispherically over an angle of 180 degrees, both vertically and horizontally.

Enclosures

Since virtually all speakers are sold as complete units (Figs. 6-11 and 6-12)—woofers and tweeters in an acoustically matched enclosure—you need not concern yourself with the details of speaker construction or the choice of a suitable speaker enclosure. The theories and opinions current on this subject would fill many books the size of this one. However, you should distinguish between two basic types of speaker enclosures—the vented and unvented.

The speaker enclosure interacts with the moving cone of the speaker, much as a violin interacts with the vibrating

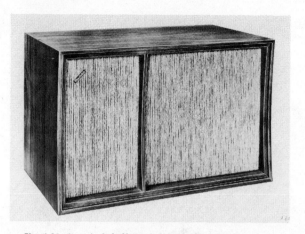

Fig. 6-11. A typical shelf-sized, full-range speaker system.

string. Many enclosures have carefully calculated openings (Fig. 6-13), or *vents*, that correspond in function to the f-holes of a violin—they control the resonance of the enclosure and help in the transfer of sonic energy to the air of the listening room. Other enclosures operate without such openings, the principle being that the enclosure remains nonresonant so as not to add its own tonal coloration to the reproduced sound. Theoretically, the resonance of the vented enclosures adds more artificial tonal coloration to the bass than the nonresonant, nonvented enclosures, but in a well-designed speaker system all resonances are so carefully controlled that the difference is largely academic.

Fig. 6-12. A furniture-style, full-sized speaker system.

Fig. 6-13. Sound path in bass-reflex type of speaker system.

Transient Response

The requirement for accurate reproduction of transients, already discussed in connection with amplifiers, applies even more stringently to speakers. To reproduce the sudden onset and cessation of sounds without blurring, the speaker cone must be able to start and stop its motion very rapidly with minimum delay due to inertia and momentum. Percussive music, full of sharp, clicking sounds and drum beats, is often used for listening tests of the transient response of a speaker. However, many audio experts feel that a more revealing test is to play some fully scored orchestral music and see if the speaker maintains a clear, transparent image of the orchestral texture without making it thick and muddy in the climaxes.

Among all these subjective considerations in speaker performance, one objective factor remains—the power rating (or power-handling capacity). The power rating tells how many watts of amplifier power the speaker can absorb. If you have a high-power amplifier capable of delivering, say, 50 watts per channel, it might be a good idea to get a speaker with a correspondingly high power rating. If you use speakers of only moderate power rating—25 watts or so—you might accidentally damage them by turning up the volume full force and applying the full 50 watts of which your amplifier is capable.

Speaker power ratings tell the amount of power the speaker can handle; they *do not* state the amount of power the speaker needs, which is naturally far less. The amount of power the speaker needs to produce full volume is known as the *power requirement*. Depending on the efficiency of the speaker, this mary vary all the way from 5 watts to 25 watts. Be sure that your amplifier output meets the power requirement of the speakers you plan to use with it. And remember that the power requirement of speakers is always stated on a per-channel basis. <u>If each of your speakers requires 20 watts, you need what is generally called a 40-watt amplifier, i.e., an amplifier delivering 20 watts per channel.</u>

BANTAM SPEAKERS

A new breed of speakers has lately gained wide popularity. They differ from their predecessors in being much smaller. Some of them, in fact, are hardly bigger than a shoebox. Surprisingly, some of these speakers attain good musical results despite their smallness. They may be slightly weak in bottom bass—say below 60 Hz—but on the whole, their tonal balance is musically adequate and pleasing.

The obvious question arises why it is now possible to attain such tonal range and quality in small speakers when it couldn't be done before. Part of the answer lies in a fresh engineering approach based on the premise that good bass does not depend solely on speaker size. What counts is the amount of air pumped out by the speaker, along with the natural resonance of the cone.

It is of course true that, other factors being equal, big speakers have the advantage in bass response. Being heavy, their cones have a naturally lower resonance; being big, they push more air. But speaker designers found that they can lower the resonance of even a small cone by adding to its weight and mounting it in a soft, floppy suspension.

The loose mounting lets the cone travel a greater distance on each swing so that it scoops up a larger quantity of air. In this way a small cone can do the work formerly accomplished only by a big one. One obvious advantage of bantam speakers —in addition to compactness—is lower price. This is the result of savings in cabinetry cost, thanks to smaller overall dimensions.

By definition, the bantam speakers represent a compromise. However they fill a legitimate need wherever space is at a premium or portability is a must. Before deciding on a bantam speaker, you should carefully compare its performance with that of several standard-sized bookshelf models in actual listening tests.

PHONO CARTRIDGES

Like speakers, phono cartridges function as transducers; that is, they change energy from one form to another. The speaker changes electricity into sound; the cartridge changes the mechanical vibrations of the stylus into a corresponding electric voltage. Because the function of the cartridge depends on mechanical motion over a wide range of frequencies, problems of resonance and tonal coloration arise. Hence, there is a marked difference in the tonal character of different cartridge designs, almost as dramatic as the difference between various makes of speakers. Some have a warm, rounded sound that seems to emphasize the orchestral blending of instruments; others have a sharper, crisper sound that seems to pick out the individual instruments rather than present them in an overall impression. Again the choice is one of personal taste.

The basic fidelity of a cartridge depends in large measure on the ability of the stylus to accurately track the intricate pattern of the record groove, for if the stylus fails to follow the exact waveform stamped into the record, the result is distortion. When you consider that tracking a stereo record groove involves a complex path full of hairpin turns to be executed at high speed—in addition to swift ups and downs—it is understandable that the mechanical demands on the stylus of a stereo cartridge are extremely severe.

An enlarged model of a stylus riding in a stereo record groove is shown in Fig. 6-14. Accurate tracing of contours is essential for undistorted reproduction. The stylus shown here has an elliptical tip contour, permitting better tracking of higher frequencies because of the thin side face.

Compliance

Most indicative among cartridge specifications are *compliance* and *dynamic mass*. Compliance tells how freely the stylus

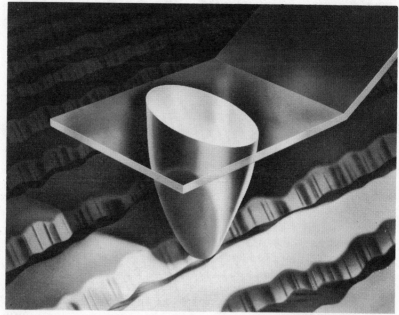

Courtesy Shure Bros., Inc.

Fig. 6-14. Enlarged model of stylus riding in stereo record groove.

moves in the groove—hence, how accurately it will follow the twists and turns. You may find compliance stated in terms like 15×10^{-6} cm/dyne. This may look forbiddingly technical, but it is quite simple to interpret as a cue to performance. Literally, the statement means that if a force of 1 dyne (about 0.000002 pound) pushes on the stylus, the stylus has enough "give" to move a distance of 15 millionths of a centimeter. But these fine physical measurements need not concern you as such. What really matters for purposes of comparison is just the first figure—the one before the multiplication sign—in this case the 15. The higher that figure, the more compliant is the cartridge. As a rough guide, consider that 15×10^{-6} cm/dyne is very adequate compliance.

When compliance gets above about 20×10^{-6} cm/dyne, the stylus becomes so pliable that it can be used satisfactorily only if the tone arm is capable of tracking at very light pressure. Cartridges with more than 20×10^{-6} cm/dyne compliance should be used only in high-quality tone arms that are carefully counterbalanced and have low-friction bearings. Putting such a cartridge in an ordinary record changer would deteriorate its performance.

Tracking Pressure

The required tracking pressure of a cartridge (i.e., the downward force necessary to make it stay in the groove) is closely related to compliance. The higher the compliance, the less weight is needed to bear down on the stylus because it more easily accepts guidance from the groove. Highly compliant cartridges (30×10^{-6} cm/dyne, or thereabouts) will track at less than a gram (about 3/100 of an ounce). At such featherweight pressures, wear on both the record and stylus is minimized. If you otherwise take good care of your records (mainly, keep them free of dust and fingerprints), a high-compliance cartridge will make them last almost indefinitely. And the stylus in the cartridge (which should be a diamond) will last for years at such light tracking pressure. But again, light tracking pressures can be achieved only in first-rate tone arms. Tone arms found on cheap changers usually require tracking pressures of 3 grams or more, rapidly accelerating both record and stylus wear. By contrast, top-quality changers, known as automatic turntables, will accommodate even the most sensitive cartridges tracking at less than 1 gram.

Dynamic Mass

Another important mechanical factor of cartridge performance is dynamic mass. Expressed in nontechnical terms this simply means the total weight of the moving parts. The lower this weight, the less the stylus will be distracted by inertia and momentum from following those fast and sharp hairpin turns that spell out the music on the record.

But if the stylus shank is too light, another problem arises: the thin metal shaft becomes flexible and the motion of the diamond tip is not accurately transferred to the electrical generating parts of the cartridge. Since the weight of the diamond is more or less fixed by its dimensions, an optimum compromise must be worked out between weight and rigidity of the shank.

Fortunately, modern metallurgy—stimulated by the requirements of space exploration—has come up with some extremely tough, lightweight alloys. Taking advantage of these materials, cartridge designers have recently been able to reduce the dynamic mass of the stylus without losing high-frequency transmission along the shaft. The dynamic mass of a modern high-performance cartridge is usually specified at 1 milligram or less.

Stylus Dimensions

The mechanical behavior of a cartridge—as distinct from its electrical properties—is also determined by the shape and size of the diamond tip. Contrary to a widespread notion, the diamond is not sharpened to a conical point like a pencil. That would rip the record, no matter how light the tracking weight. Rather, the tip is rounded, and the radius of its curvature varies among different cartridge models. Most manufacturers offer a choice of 0.7-, 0.5-, and 0.3-mil styli, 1 mil being equal to 1/1000th of an inch.

The smaller styli are capable of clearer treble reproduction because they fit more snugly into the tight little curves—especially near the center of the record where the musical waveforms are more densely packed together in the groove. However, the 0.5- and 0.7-mil styli tend to rattle loosely in some of the older monophonic records which were cut out with wider grooves. The 0.5-mil stylus tracks both new stereo and old mono records quite adequately and can be recommended as a universal stylus for any LP—mono or stereo—regardless of age.

Most of the better cartridges today feature a stylus with an elliptical profile, which tracks the record groove more accurately at high frequencies. Such styli usually combine a width of 0.7 mil with an end radius of 0.2 mil, which enables the narrow part of the stylus to fit more snugly between the tightly spaced wiggles of the record groove. A recent innovation in stylus design is the so-called Shibata stylus (named after its Japanese inventor) which is featured on the most advanced and expensive cartridge models. Here the diamond tip is cut with specially shaped facets to increase the contact area between the stylus and the groove for more accurate guidance of the stylus motion by the groove wall.

Frequency Response and Resonance

Making the moving parts of a cartridge so extremely light also results in advantages with respect to frequency response. The lightness of the moving parts raises their natural resonance beyond the audible range, usually above 20,000 hertz. Consequently, harsh-sounding resonance peaks are avoided within the audible spectrum, and the overall range of the cartridge can be smoothly extended. This largely accounts for the great improvement in high-frequency sound achieved in cartridge design within the last few years, eliminating the stridency that had previously been associated with extended range. The frequency response of a good stereo cartridge should extend to 20,000 Hz or beyond, and the maximum deviation from flat response should not exceed ±3 dB.

Stereo Separation

Separate signals are generated within the stereo cartridge for left and right channels—a feat accomplished by having two separate voltage generators arranged at an angle within the cartridge to distinguish the lateral and vertical motions of the stylus. It is vital for effective stereo to keep the two signals properly separated, preventing the left-channel signal from leaking over to the right, and vice versa—a difficulty known as "cross talk." At the customary measuring point of 1,000 hertz, a good cartridge should have at least 20-dB separation, meaning that any left-channel signal appearing in the right channel should be at least 20 dB less loud than the proper signal belonging to that channel. And the same should hold true the other way round. Separation has a tendency to diminish at higher frequencies, with greater chance of leakage between the channels. That is why, for many of the better cartridges, separation is also specified at some higher point in the frequency range. A good cartridge should still maintain about 15-dB separation all the way up to 10,000 hertz.

Cartridge Types

Basically, phonograph cartridges come in two types—ceramic and magnetic. Virtually all high-fidelity cartridges operate on the magnetic principle, resembling, in their interior structure, a miniature power plant. In some designs small magnets linked to the stylus move inside tiny coils, generating voltages proportional to the stylus motion. These are known as moving-magnet cartridges. In other designs the process is reversed—the coils move and the magnet remains fixed. Quite logically, these are known as moving-coil cartridges. A third type, utilizing the motion of a neutral iron shank in a magnetic field, is called a moving-iron cartridge. Magnetic types are essentially similar in performance and differ only in their mechanical structure. In output level (3 to 8 millivolts), in frequency characteristics, and in a rather technical factor known as load impedance, all these cartridges match the standard "magnetic phono" inputs found at the rear of modern amplifiers or preamplifiers. All you need to do is to plug in the pin plugs on the phono cartridge leads and the unit is ready for operation.

A final caution: the cartridge is an extremely delicate device, and a highly compliant stylus is easily damaged by rough handling. Treat it with care and set it very gently on the record.

TURNTABLES

While all other components of a stereo system are directly involved—either mechanically or electrically—with the actual sound to be reproduced, the turntable and tone arm stay mute. They are the silent partners of the team. No sound—either in mechanical or electrical form—passes through them; but like most silent partners, they have considerable influence on the whole operation of the system.

Rumble

Staying silent is by no means easy for a turntable. Like any rotating device, it is prone to vibration. Such vibration is

Courtesy Thorens Co.

Fig. 6-15. The underside of a typical high-fidelity phono turntable.

Courtesy United Audio Product, Inc.

Fig. 6-16. A top-quality record changer matches the quality of professional-type turntable and tone arms.

picked up by the phono cartridge along with the music, causing the notorious *turntable rumble*—an unwelcome accompaniment to the music that sounds like softly rolling thunder in the distance. Especially if you have speakers with excellent bass response, rumble, which usually consists of the lowest frequencies in the audible spectrum, can be quite annoying.

The only way to eliminate rumble is by means of smooth-running turntable motors and precision-machined and carefully fitted bearings and parts. This requires painstaking manufacturing and inspection methods and is the reason why good turntables, despite their essential simplicity, tend to be relatively expensive. Figs. 6-15 and 6-16 show the construction of a typical turntable.

Aside from keeping all mechanical tolerances close, vibration is further reduced by suspending the drive motor in elastic shock mounts. Moreover, motor vibration is isolated from the turning platter by an elastic transmission (usually in the form of a plastic belt) that filters out whatever residual vibration remains.

Turntable rumble is stated in specifications as the number of decibels of the rumbling noise as compared to a standard test tone played on a record. The minimum requirement for high fidelity, as defined by the National Association of Broadcasters, is −35 dB, meaning that the rumble must be 35 dB less in loudness than the test tone. On a speaker system with very good bass response, this could still be slightly audible. A turntable with −45-dB rumble would seem almost completely silent, and some of the best turntables attain even lower rumble ratings.

Most test records have a band of silent grooves for convenient rumble tests. As you play these grooves, be sure your tone controls are in "flat" position, i.e., without bass boost or attenuation. Even the best turntable will rumble if you turn up the bass boost too far at high volume.

Wow and Flutter

If the turntable motor won't pull evenly, the result is a chugging motion causing a fast wavering of pitch in the reproduced music. This is known as *flutter* and imparts a quivering quality to the sound, especially on long-held notes that

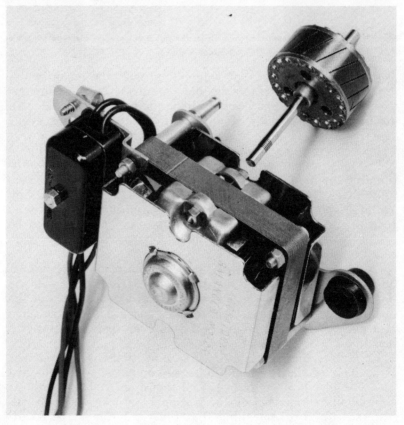

Fig. 6-17. A precision-balanced, smooth-running phono motor.

should sound steady. A slower up-and-down pitch variation, reminiscent of the wobbly sound of sirens, is descriptively known as *wow*.

In quality turntables, flutter and wow are avoided by using motors with highly uniform pull, or *torque*, if you prefer the technical term for rotary force. Some of the best turntables employ so-called hysteresis-synchronous motors shown in Fig. 6-17, whose torque remains virtually constant throughout each turn. Moreover their speed is unaffected within wide limits by changes in the supply voltage—a point to consider if you live in an area where the voltage is likely to vary. However, the less expensive four-pole motors are also capable of excellent results; if their moving parts are properly balanced and carefully assembled, their motion is quite smooth. The flywheel effect of the turntable platter itself contributes to smoothness of rotation. For this reason, quality turntable platters are carefully machined, and they are balanced just as the wheels on your car are balanced for even rotation.

Wow and flutter are expressed as a percentage. On very good turntables wow and flutter might be about 0.1%, which is inaudible. One informal test is to play a record with long-sustained piano chords and observe whether the tone remains steady. Such sustained chords show up pitch variations very clearly. Be sure, however, that the waver, if there is any, is not on the record itself; try different records. Most test records also have constant tones recorded on them by which such wow and flutter tests can be easily performed.

Speed Accuracy

Another important consideration is true speed—whether the turntable actually turns at the proper rate. Otherwise, the pitch throughout the record will be either too high or too low. Turntable speed should be accurate within 1%, and people with an especially critical sense of pitch might find 0.5% a more satisfactory standard. If you have a sense of absolute pitch (very few people do), or if you are a musician who likes to play his instrument along with the recorded music, you may prefer a turntable with a variable-speed adjustment. This permits you, within limits, to "tune" the turntable so that it will be on pitch for your instrument. This control may also be used to compensate for speed inaccuracies due to variations in your supply of electricity. A stroboscope disc placed on the turntable provides a simple way of checking speed accuracy, and many turntables have such a disc built in.

Two philosophies are evident in current turntable design. One might be called the battleship approach—relying on heavy, massive construction to increase the flywheel effect and to achieve smooth rotation. The other approach is exactly opposite, emphasizing lightness as a way to keep vibrational movements small enough to be effectively controlled. Both methods have led to excellent results when followed by capable designers, so don't judge a turntable by weight alone.

You will have to decide whether to get a single-speed or a multispeed turntable. Unless you still have some old 78-rpm discs you like to play occasionally, you will hardly find any need for a speed other than 33⅓ rpm, which is standard for both mono and stereo albums. Since the speed-shift mechanism on multispeed turntables usually increases the price considerably, confining yourself to 33⅓ rpm is likely to save you a fair amount of money and may thus let you buy a better turntable for your budget. Almost anything worthwhile recorded on 45-rpm discs is also available on 33⅓-rpm records. Automatic turntables usually offer a choice of three speeds (33⅓, 45, and 78 rpm) as well as a choice of either manual or automatic operation.

TONE ARMS

Almost any phono cartridge will sound better if given a chance by a good tone arm; the better the cartridge, the greater is the improvement attainable through the arm. Moreover, a good arm permits lighter tracking pressure, thus saving wear on both records and stylus.

At first glance the job of the tone arm seems utterly simple—just to hold the cartridge in place. But in fact, several precise considerations are involved in the proper performance of this simple function. The tone arm must guide the cartridge across the disc in an intricately calculated path; it must minimize all friction drag, maintain correct downward pressure regardless of record warp, floor vibration, and other distracting factors, and isolate the cartridge from resonant vibration so that it responds only to the undulations of the record groove.

Tracking Error

The first requirement is to keep the cartridge so lined up with the record grooves as to minimize the effect of changing groove diameter as the arm moves across the disc. Ideally, the cartridge should always remain tangent to the groove; practically, this is impossible because the arm does not move in a straight radial line across the record, but turns on a pivot. The geometry of a well-designed arm—its curves and angles relative to its pivotal point—reduces the departure from the true tangent position, which is called the *tracking error* (Fig. 6-18). This assures that the stylus does not ride athwart the record groove and responds equally to both groove walls.

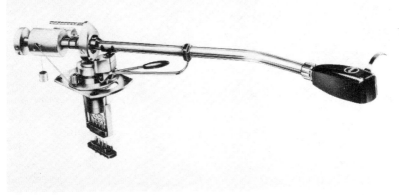

Courtesy Shure Bros., Inc.

Fig. 6-18. A professional-quality tone arm with low-friction gimbal bearings, antiskating compensator, and cueing device.

Friction Drag

The second requirement, low friction, is accomplished in various ways in different designs. Some arms rest on a single needle point to allow free motion; others employ miniature ball bearings; still others use gimbal suspensions with pinion bearings or knife-edge pivots like those used in chemical balances. Again, this calls for precision manufacture to very close tolerances.

Tracking Pressure

Various methods have been employed to provide constant stylus pressure (Fig. 6-19). Ordinary arms usually rely on counterbalancing the tone-arm weight by upward-directed spring tension. This is inherently unstable, and the upward motion of a warped disc or floor vibration reinforces the upward pull so that contact between stylus and groove becomes erratic and intermittent. At worst, it can make the stylus jump grooves. Quality tone arms therefore are balanced not by upward spring tension but by a stable, adjustable counterweight.

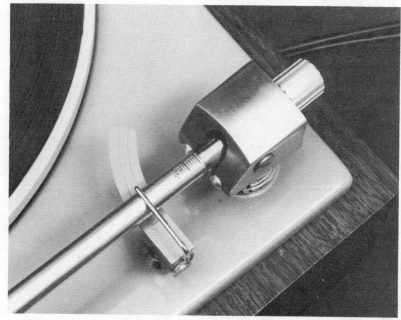

Courtesy Bogen-Presto Div., Siegler Corp.

Fig. 6-19. Tone arm with calibrated scale for adjustment of stylus pressure.

Many arms employ a system known as dynamic balance. This also uses spring tension, but the direction of the pull is downward rather than upward. To keep the downward pull from being too heavy, the arm is also counterbalanced by a static counterweight. The spring applies a constant force that keeps the arm tracking at the right pressure, regardless of external accelerations imposed by record warp or floor vibration. In this way it maintains reliable contact with both sides of the record groove, largely independent of outward influence. Moreover, the pressure is applied evenly to both walls, even if the turntable is not strictly level. In houses with slanting floors this is a great convenience, and even jolts transmitted to the turntable through a shaky floor will not jerk the arm from the groove.

Resonance

The natural resonant frequency of the arm itself must be so low that the arm will not vibrate at any of the musical frequencies encountered by the cartridge in the groove. With a resonance below audible frequencies, no arm vibration will distract the cartridge as it traces the bass vibration of the music.

Because high-compliance cartridges are extremely sensitive to rough handling, some tone arms are now equipped with special safety devices to prevent cartridge damage. Some have a viscous-damping fluid in their bearings which lets the arm float gently down toward the record if the arm is accidentally dropped. Others have automatic positioning devices which lower the arm gently on the disc, protecting both the cartridge and the record.

Skating Compensation

A recent innovation in top-grade tone arms is antiskating devices. The term skating, as used in audio, has nothing to do with ice or the rink. Rather, the word refers to a kind of inverse centrifugal force that tends to push the tone arm toward the center of the record, even if the turntable is perfectly level. The invisible force that grabs the arm and pushes on it is caused by a tricky combination of factors—mostly the drag of stylus friction acting against the odd angle at which the arm scans the disc. The net result is a constant nudge on the arm.

In the past, this slight imbalance was simply ignored. Engineers got away with this because most older cartridges were so stiff jointed that a little sideways push did not faze them much. But today's best high-compliance cartridges are riding the grooves at less than a gram pressure and are far more sensitive to even a slight imbalance.

To visualize what happens, picture the sylus nestled in the groove of the stereo disc. Ideally, it should contact both sides of the groove with equal pressure so that both left and right stereo channels are precisely traced. Yet the skating force shoves the stylus up against the inner-groove wall. The result is triple trouble:

1. The extra push on the inner wall adds extra wear to that side; also the inner stylus surface wears sooner.

2. Pushed away from the outer groove wall, the stylus no longer makes permanent contact with all the wiggles. For brief moments, it rattles loosely in the groove and no longer follows the musical waveform. The result is distortion in one channel—the one represented by the outer wall.

3. One sided pressure constantly acting on the stylus pushes it off center. This may cause a permanent kink in the sty-

lus mechanism and unbalance the stereo effect by limiting free stylus swing to one side.

For this reason, antiskating devices have been incorporated in many modern tone arms, including those of the better types of record changers. These antiskating devices differ in various models. Some operate through small weights pulling on the arm, others through spring tension. Yet they are all similar in principle. They apply force to the tone arm equal and opposite to the skating force so as to cancel out the imbalance.

RECORD CHANGERS

Record changers are, in effect, combinations of tone arms and turntables with built-in automation. The same requirements apply as for separate turntables and tone arms. It should be noted that many currently available changers do not meet true high-fidelity standards. The term automatic turntable has been coined for changers that combined precision turntables with high-quality tone arms.

The best of these equal the performance of so-called professional turntables and arms and offer the added advantage of automatic positioning of the tone arm at the beginning of each record and automatic shutoff at the end. The fact that automatic turntables eliminate the need for handling the tone arm is an added safety factor in families where fumble-fingered persons might accidentally drop the tone arm on the record or otherwise damage the delicate cartridge by rough treatment.

TAPE RECORDERS

The quality requirements for high-fidelity tape recorders are essentially consistent with corresponding specifications for the other components discussed. A good recorder should have a frequency response from 30 to 18,000 Hz or beyond (at a speed of 7½ ips) with a deviation from flat response no greater than ±3 dB. Many inferior recorders have response peaks of 5 or 6 dB or more within this range, which definitely contributes unnatural coloration or harshness to the recorded sound. The signal-to-noise ratio (hum) should be better than 45 dB. An acceptable limit for wow and flutter is about 0.3%. A tape machine meeting these specifications provides sound quality on a par with other high-performance sound components.

Open Reel Machines

Tape recorders differ widely in operating features and mechanical design, and in that respect no general rules can be set down. Many of the better open reel machines use three separate drive motors—one for each reel and one for the critical capstan drive that pulls the tape past the magnetic heads. Yet there are excellent machines using only a single motor for all three functions. Similarly, most quality machines employ three separate heads for recording, playback, and erasing; yet some perfectly good machines combine record and playback functions in a single head, though that inherently entails a certain compromise in frequency response and does not permit monitoring directly from the tape while the recording is being made.

The best way to judge the mechanical performance of a tape recorder is to put it through its paces, working the various controls and observing whether the tape starts and stops smoothly, without jerking, and whether the machine runs without excessive vibration or obtrusive whirring and other distracting mechanical noises. Switching back and forth between rewind and fast-forward speeds quickly reveals any possible tendencies to break or spill the tape—disastrous events that should never happen on a good machine.

A stereo recorder should be equipped for four-track operation—two tracks in each direction of tape travel—with optional facilities for recording monaurally on all four tracks (Fig. 6-20). Standard speeds are 7½ and 3¾ inches per second (ips). Some recorders offer a third speed of 1⅞ ips, which is adequate for recording speech, although frequency response at such slow speeds does not permit high-fidelity recording of music. The purpose of the extra-slow speed is to permit highly economical speech recording—eight hours or so (mono) on a single standard reel of tape.

Of course, the recorder should have the proper input and output facilities to be connected permanently into your high-fidelity system, and the dealer demonstrating the tape machine should explain these details to you along with the various operating features that differ widely among individual designs.

For a very reliable test of the overall performance of a tape recorder, take a good stereo record and copy it on tape. Then play back the tape and compare the sound to that of the original recording. On a good tape machine there should be no appreciable difference.

Courtesy Astrocom/Marlux Inc.

Fig. 6-20. A high-quality stereo tape recorder.

Among the operating features of tape recorders, one deserves special mention: automatic reversal. This feature has been incorporated into several models in the quality price bracket. The innovation is of particular interest for those who like to record radio programs. Most amateur recordists have experienced agonized moments when the duration of a broadcast they are taping exceeds the playing time of the tape on the supply reel. And nothing is more frustrating than to lose a crucial passage while frantically trying to switch reels in midstream. Self-reversing tape recorders are designed to put an end to such mishaps.

In normal recording or playback, tape travels from the left reel to the right. If you come to the end of a reel (having recorded one track), you flip over the reel and shift the full take-up reel from the right to the left hub so that left-to-right recording can continue. The flip-over may take about 30 seconds even for a nimble operator—and that slice of the broadcast is irretrievably lost. That's why commercial studios always run two tape machines in tandem.

The self-reversing machines let you skip this desperate maneuver. They are able to record and play with the tape running either left to right or right to left. When you come to the end of a reel, just flick a switch and the recording continues in the opposite direction on another track. Only a second or so is lost in reversal. Some self-reversing models are fully automatic. You do not need to be there to flick the switch as the reel runs out. But in any case, make sure that the machine you pick has tape reversal for recording. Some machines have this feature only for playback.

Tape Cassettes and Cartridges

Fig. 6-21 shows the three basic type formats—open reel, 8-track cartridge, and cassette. Open-reel tape still offers the best fidelity, but cassette recorders have been perfected to the point where they closely approach the quality of open-reel machines. In addition, cassettes offer the advantage of compactness and simplicity of operation. This accounts for their growing popularity. In cassette machines, all you do is insert the cassette, push the button, and you're ready to roll. Open-reel machines, by contrast, require tape threading.

Courtesy Ampex, Consumer Equipment Div.

Fig. 6-21. The three basic tape formats.

Not all cassette machines attain adequate fidelity. Low-priced, battery-powered portable cassette machines make no pretense to tonal quality. But there are many high-performance cassette decks built to precision standards. These decks are plugged into a stereo system exactly the way a standard tape deck is plugged in, playing through the amplifier and the speakers of the main system. This is the only kind of cassette equipment to be considered for high-fidelity use. A good cassette deck of this type offers a frequency range to about 12,000 Hz, which yields very satisfying sound, just a shade below the best disc recordings or open-reel tape.

The operating features of cassette decks are similar to those of standard open-reel recorders. Some of the high-quality cassette decks also feature automatic reverse for long periods of nearly uninterrupted recording and playback. In terms of printed specifications, the quality factors of cassette decks are the same as for reel recorders. Because the tape speed in cassette machines is very slow (a mere 1⅞ inches per second), flutter and wow is especially noticeable on poor cassette machines. On a good model, flutter and wow should not exceed 0.2%.

Some of the better cassette decks feature the so-called Dolby circuit, so named after its British inventor. The purpose of this device is to reduce the hissing background noise that comes from the tape itself, thereby allowing very soft passages of music to stand out more clearly. As a rule, this feature adds at least $50 to the cost of the deck. For the perfectionist intent on making near-perfect cassette recordings it is well worth the price. However, the problem of background noise can also be tackled by buying high-quality cassettes in which special low-noise tape is used. With those cassettes, the background hiss is so slight that many listeners do not find it objectionable. Consequently, they will not find the Dolby feature necessary for their enjoyment of the music, and by buying a cassette deck without the built-in Dolby they can save a considerable amount of money.

Plug-in decks are also available for the 8-track cartridges that have gained wide acceptance in automobile stereo systems. They are convenient for playing stereo and four-channel cartridges, through home-based audio systems. Some of these decks also have facilities for making your own recordings on 8-track cartridges. The unit pictured in Fig. 6-22 can be used for recording and playback and has a built-in Dolby noise suppressor.

Courtesy 3M/Wollensak
Fig. 6-22. A high-quality tape cartridge deck.

EARPHONES

Among listeners who like to shut out the distractions in their homes, earphones have recently gained much popularity. They offer instant privacy in listening. And this privacy works both ways: it insulates you from outside noises and, at the same time, keeps your music from bothering other people in the house.

Those who think that earphones went out with "catwhisker" radios would hardly recognize today's stereo and quadraphonic models. In the early years of radio, earphones were nothing but primitive telephone receivers, hardly capable of even the most rudimentary fidelity. By contrast, modern earphones —aside from being two- or four-channel devices—are crafted with the same precision that marks other quality audio components. Structurally they resemble modern speakers, consisting essentially of a permanent magnet, a voice coil, and a carefully designed and suitably suspended cone. By way of analogy, one might say that today's stereo headsets compare to ordinary radio earphones as a racing yacht compares to a tree-trunk canoe. Mainly the difference lies in calculated refinement.

Cone and magnet structures of the new headsets are designed for extended frequency response and low distortion. Many current models cover the range of about 30 to 20,000 hertz with less than 1% distortion at maximum output—specifications that few full-sized speakers can match. If you have had no prior experience with modern headsets, you will be amazed at the full bass obtained from such small sound generators. How can low frequencies be so effectively reproduced by diaphragms measuring only about two inches in diameter? Ordinary speakers must be relatively large and for adequate bass response they must push plenty of air to project bass energy into a room-sized listening space. But the "listening space" to be filled by earphone is only the tiny air volume between the earphones and your eardrums. Moreover, with the headsets fitted tightly against your ears by means of soft padding, this small air space is sealed off and represents what engineers call a "closed system." This provides practically loss-free transfer of low-frequency energy. Under such conditions, even a small sound generator suffices to create ample bass.

Like speakers, stereo headsets tend to have their individual sound coloration. In selecting a model for your own use, compare different makes just as if you were buying a pair of speakers. Clarity is the most important criterion. Make sure the sound does not blur, even at full volume, and that the violins sound silky and smooth, without stridency.

Make sure the earphones fit you properly, so you can wear them all evening long without discomfort. Fit around the head is rarely a problem because most headbands are either flexible or otherwise adjustable. The earpieces, however, have fixed

Fig. 6-23. Modern stereo earphones of lightweight open-air design.

Courtesy Sennheiser Electronic Corp.

dimensions. So make sure that they do not pinch or squeeze your ears. They should fit *around* the ears rather than over them.

More recently, a new type of earphone (Fig. 6-23) has been introduced which does not seal off the ear. Instead, these earphones use foam rubber pads that permit airflow between the ear and the surroundings. Such earphones are especially comfortable even for prolonged listening and they also tend to be much lighter than the conventional kind. The sound quality of the better headphones of this type is especially good, providing a very natural, spacious sound. Before deciding whether to buy a closed-shell earphone or one of the open-air foam-cushioned types, you should make your own personal comparisons.

Hooking up the earphones presents no difficulties with modern equipment. Nearly all up-to-date-amplifiers have special stereo earphone jacks.

CHAPTER 7

Sound Value for Your Dollar

The preceding chapter attempted to provide a numerical definition of high fidelity by putting facts in figures. The discussion now turns to numbers of a different sort, dollars and cents, hoping to establish a sensible relation between price and performance.

The price range of high-fidelity components is extremely wide. A single component in the deluxe category may cost as much as a whole sound system in a more modest price class. Considering this enormous price spread, some perspective is needed to reconcile your aspirations and your budget. The trick is to decide at which point in the price scale your investment and your interest are best rewarded.

Of course, the decision depends both on the state of your personal finances and the degree of your passion for good sound. To help guide your choice, consider three price groups and what you can expect for your money at each of these three levels:

1) A "bottom dollar" stereo system
2) A "golden medium" stereo system
3) A "deluxe" stereo system

"BOTTOM DOLLAR" STEREO

The lowest price of admission to component stereo is somewhere around $300. A good turntable-tone-arm combination can be had for around $100, and you can fit it with a good cartridge in the price range from about $25 to $35. For approximately $120 you can get an amplifier with moderate power output and a number of good-sounding but inexpensive speaker systems can be had for about $50-70. Even below this price range it is possible to find some fine-sounding speakers, but they are the exception rather than the rule. If shopping for low-cost speakers, check them out carefully for properly balanced sound, avoiding those with thumpy bass, and especially those with overly bright, metallic highs.

You can save money by buying your speakers in unfinished cabinets rather than in walnut or other expensive finishes. By applying wood stain you can easily get the unfinished cabinets to match your furniture.

To stay within your budget you may start with a basic stereo record-playing system and add an fm tuner or am/fm receiver later when your finances have recovered from the first outlay. Prices for stereo fm tuners start at about $80. Electronic mail-order firms offer attractive bargains in this price range, and it may be a good idea to study their catalogs if you are considering equipment in this class. Typical components for such a system are shown in Fig. 7-1.

What kind of performance does such a system offer? Quite likely, the lowest reaches of the bass may not come through with powerful conviction, and the limited power reserve won't let you shake the walls. But in normal-size rooms and at normal volume levels there will be no noticeable deterrents to your musical pleasure. The sound will be clean and largely undistorted; if your speakers are well chosen, the overall quality will be balanced, pleasing, and natural. The improvement in comparison to ordinary phonographs will amaze you. Listening to this system, you will not be aware of any deficiency, except when you play it in direct comparison (side by side) with the more expensive systems to be described next.

"GOLDEN MEDIUM" STEREO

An investment of about $600 puts you solidly in the middle class as far as audio is concerned. This assures listening without compromise on quality and no skimping on power. In this class, your manual or automatic turntable might cost about

Courtesy Heath Co.
(A) Turntable and tone-arm combination.

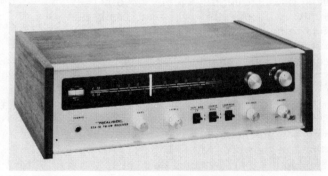

Courtesy Radio Shack, a Tandy Corp. Co.
(B) Low cost am/fm stereo receiver.

Courtesy Heath Co.
(C) Low cost bookshelf speaker assembly.

Fig. 7-1. Components of a typical "bottom-dollar" system.

$130-$150 and your cartridge between $35 and $45. With an amplifier that delivers a minimum of 25 watts per channel and an ample choice of speakers in the $80-140 range, there is nothing "middling" about this "golden medium" as far as performance is concerned. The amplifier can handle even the most demanding orchestral passages, and the speakers will take the whole musical frequency spectrum easily in their stride. The residue level of hum will be virtually imperceptible, permitting sound reproduction without undue background noise. An fm tuner in the $130 to $150 price range would be sensitive enough not to present any problems in most reception areas. At this price level your amplifier should offer ample control facilities, such as controls for remote speakers on your porch or in your bedroom, tape monitor controls, loudness compensation, blend control, etc., that make the system flexible in use and adaptable to any need.

Most listeners find themselves completely satisfied by equipment in this class. Chances are that you will find yourself among this majority—fully enjoying music without even being aware of the equipment. This is as it should be, for the ultimate test of a good sound system is that it is unobtrusive—that is, it doesn't get between you and the music. A system of his kind should pass such a critical listening test. And if you feel that your musical demands are satisfied at this level of performance, there is no reason why you should spend more. Of course, you may want to add a tape recorder—an item not included in the estimate—and this would raise your total outlay. Components of a typical "golden medium" system are shown in Fig. 7-2.

"DELUXE" STEREO

There is virtually no limit to the amount of money you can spend on an audio system. However, in general it can be said that after a certain point is reached, small improvements in reproduction require large investments in equipment.

You can spend more than $200 for a top-notch manual or automatic turntable and an amplifier rated at 60 watts per channel or more (especially if it consists of separate power amplifier and preamplifier) would run you well over $500, and you would pay as much for a top-rate tuner. As for speakers, you can get a first-rate pair for about $450, but you can pay more than $1,000 a pair for some of the more unusual and truly outstanding designs.

Compared with the "golden medium" system, the "deluxe" system may bring you only marginal gains in sonic realism,

Courtesy United Audio Products, Inc.
(A) Automatic turntable with precision tone arm and cueing control.

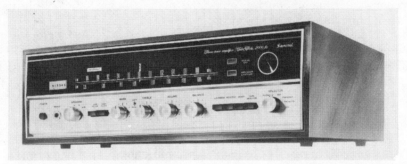

Courtesy Sansui Electronics Corp.
(B) An integrated receiver of medium power rating.

Courtesy Electric and Musical Industries, Ltd.
(C) Bookshelf speaker.
Fig. 7-2. Components for a typical "golden medium" stereo system.

but only you can decide whether such gains are worth the costs involved. Typical components in the "deluxe" class are shown in Fig. 7-3.

FOUR-CHANNEL BUDGETING

If you contemplate a four-channel system, your budget should reflect the higher cost of a quadraphonic receiver, plus the fact that you need four speakers instead of two. Figure that a "middle class" four-channel receiver may cost about $450 while a deluxe four-channel receiver may cost upward of $600. Low-cost quad receivers can be had for about $300 or less, but most units in that price class do not have sufficient power per channel to assure musical realism in heavy passages. If you recall our brief survey of the economics of four-channel sound in Chapter 3, you will understand why inexpensive four-channel equipment often has to compromise fidelity. The deluxe four-channel receiver in Fig. 6-4 has decoding circuits for all types of quadraphonic records (matrix and discrete, CD-4). It features a built-in oscilloscope for visual display of the sound distribution in the four channels.

One bright aspect in budgeting for quad is the fact that not all four speakers have to be of equal quality. Since most of the sound emanates from the two front speakers, you may use less expensive speakers with less power capacity for the two rear channels.

If you plan to play mainly matrix-type four-channel records, any standard stereo phono cartridge will do the job. But if you plan to play the so-called "discrete" CD-4 type of quadraphonic records, you will need a special CD-4 cartridge in your record player, which may cost anywhere from about $80 to $120. •

Open-reel tape recorders equipped for 4-channel operation also tend to be more expensive than standard stereo models, and you will have to budget at least $300 for a good quadraphonic tape machine.

WHERE EXTRA DOLLARS REALLY COUNT

For most of us, however, such a system is merely a daydream. It then becomes important to decide just where you might skimp a bit without seriously compromising the end result; or, conversely, where a few extra dollars would bring the biggest improvement.

Any extra money you can afford beyond your original budget is usually best invested by upgrading your speakers. Speak-

Courtesy Lenco Corp.
(A) Precision turntable/tone arm combination with cueing control.

Courtesy Kenwood Corp.
(B) Separate am/fm tuner and amplifier rated at 150 watts per channel.

(C) Top line loudspeaker with angled front for wide-angle treble distribution.

Courtesy Audio Dynamics Corp.

Fig. 7-3. Components for a typical deluxe stereo system.

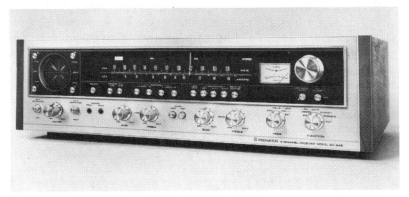

Courtesy U.S. Pioneer Electronics Corp.

Fig. 7-4. A deluxe four-channel receiver.

ers are, in effect, the voice of your sound system. What you actually hear depends largely on them. Even a modest amplifier is capable of producing a good signal within its power limits, but the qualities of such a signal are ruined or lowered before reaching your ear if the speakers are the limiting factor in your chain of components. In short, even an inexpensive amplifier is likely to sound a great deal better if you don't stint yourself on speakers.

This does not mean that you have to plunk down upward of $300 for a pair of speakers alone. For about $80 to $100 you can buy a speaker of quite excellent musical quality. Again, critical listening (as suggested in the last chapter) is the only way to tell which make and model most appeals to your ears. If you have an economy system with little power reserve, your speakers should be of fairly efficient design so that they will not require a great deal of wattage to drive them.

One surefire way to save cash in any price range is to buy a receiver instead of a separate amplifier and tuner. Because in a receiver both these components are combined on the same chassis and share the same power supply, this kind of component offers inherent cost economy. On the average, a saving of about 20% can be gained by buying a receiver in place of separate components of comparable performance.

THE PRICE OF POWER

The price of amplifiers rises rather steeply with increased power rating. This raises the question as to how many watts you really need; or, why pay a premium for superfluous watts?

Strictly speaking, extra watts are not really superfluous. As was explained in the preceding chapter, ample power reserve contributes a margin of naturalness and transparency to the sound of heavily scored passages. What must now be determined is just what constitutes adequate power reserve for your particular needs.

Again, assume three typical situations, hoping that you will find yourself right at home in one of them.

Case 1: Your living room is furnished so that its surfaces reflect rather than absorb sound. This fact alone cuts down your power requirements considerably. Smooth walls, no heavy curtains, tile or linoleum floor, and no heavy upholstery—all these contribute to a "live" acoustic environment and make sound louder, even at very moderate power. In such surroundings 10 to 15 watts per channel usually suffices.

Case 2: A more typical situation is a room with curtains and some rugs (though not covering the entire floor area), an uncluttered furniture arrangement and not too much heavy upholstery. For this arrangement about 20 to 25 watts per channel is usually sufficient.

Case 3: Here is what is called an acoustically "dead" environment. Wall-to-wall carpeting, heavy draperies, stuffed chairs, couches, pillows, and wall hangings—all soak up sound, and extra amplifier power is needed to make up for it. In this type of room you will need from 25 to 50 watts per channel, or even more.

These figures are based on the assumption that you like to play symphonic orchestrations at full volume. If your listening tastes run to chamber music or jazz combos, which make considerably lighter power demands on your equipment, you will not need power ratings this large.

Another assumption is that your listening room is of average size, not exceeding 30 feet in either length or width, and with a ceiling of normal height. In a large room you may have to increase the suggested power ratings.

A STRATEGY FOR SHOPPING

In conclusion, here are just a few suggestions to make your hi-fi shopping more efficient, as well as more enjoyable. Before you enter the store, have a pretty firm idea about your budget. Tell the dealer right away what your budget is so that both of you will be thinking along the same lines. A competent dealer will see to it that you get the best sound for your dollar. No doubt he will try to persuade you to upgrade your choice. Make him prove his point. Let him demonstrate the extra features you would get for spending a little more. Compare the sound by switching back and forth between a system within your budget and one slightly beyond. Then decide whether the difference in sound is worth the difference in cost. And just in case you decide that it is, leave about $100 leeway on your budget.

When you are comparing different components, compare just two at a time. Trying to make comparisons among three or more different items simultaneously is only going to confuse you. So make your choices one-by-one in two-at-a-time comparisons.

Naturally, this takes time. Hurrying through your shopping routine puts heavy odds against your ultimate satisfaction. It's not a good idea, therefore, to ask for an equipment demonstration during peak shopping hours when the store is crowded and the salesman can give you only limited attention. It might be best to make an appointment for a time when you would be able to listen and talk with less distraction; also, you might ask the salesman to hook up in advance any components in which you are particularly interested.

CHAPTER 8

Kits for Cash and Pleasure

If you are a do-it-yourselfer, high-fidelity kits promise double gain: money saving and the satisfaction of having built your equipment with your own hands. Since labor is one of the most expensive ingredients of amplifiers and tuners, you can save an average of 30 to 40% by assembling your own components. With these savings you may be able to afford equipment that otherwise would be beyond your means.

Until a few years ago, kit building used to be a haphazard venture to be recommended only to the technically adept. All this has drastically changed. Kits now have appearance comparable to factory-made components. Many manufacturers offer the option of buying their outstanding models either as kits or in factory-finished form. Those kits, except for final assembly, are identical in every respect to the standard models.

THE PRACTICALLY FOOLPROOF KIT

The most important development in kits has been ingenious ways to reduce the chance of error on the part of the kit builder. One such innovation is the use of modules (Fig. 8-1). The step-by-step instructions now furnished with kits simply won't let you make a mistake, providing you follow the instruc-

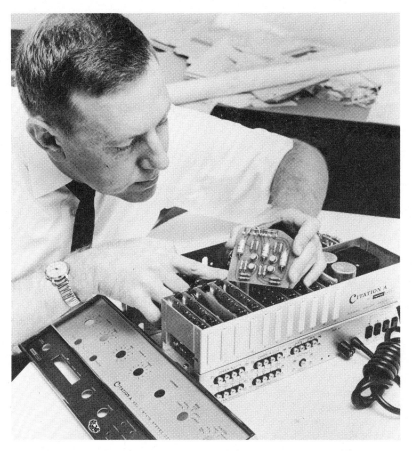

Fig. 8-1. A transistor preamplifier containing modules that are built separately and then slipped in place in the amplifier.

tions carefully. And just in case you slip up anyway, a double-checking routine that is part of most assembly procedures immediately lets you catch your own mistake.

Another improvement in kit design is the package itself (Fig. 8-2). No longer are the parts thrown together at random, leaving you to fish and hunt for every item. In most kits the various types of component parts—capacitors, resistors, etc.—are presorted for you.

In some of the newer kits the package itself serves as the work-bench. The bottom folds out to form your work area while the lid serves as storage space from which the parts are withdrawn as you use them. If, like many apartment dwellers,

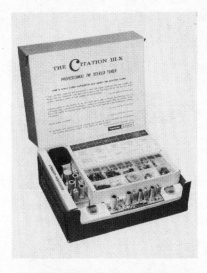

Fig. 8-2. An example of kit packaging.

you lack proper workspace for do-it-yourself ventures, these packages are a decided convenience and help keep your kit project neat and unconfused from beginning to end.

SHOULD YOU BUY OR BUILD IT?

While no special skills are required for kit building, a certain amount of basic handiness is definitely helpful. If you are normally adept, you will not encounter serious problems. True, previous experience is helpful, but it is by no means indispensable. All necessary techniques are fully explained in the instructions, including the all-important knack of making proper solder joints. And if you have a friend who can show you how it's done, so much the better. (Caution: always use rosin-core solder for electronic work; never use acid-core solder.)

Far more important than mechanical skill are certain character traits. Are you able to sit still for a fairly long time, giving concentrated attention to the task at hand? Are you methodical in your work habits—able to follow instructions patiently without instinctively seeking a short cut? Answer these questions honestly before you undertake a kit project. They are a kit builder's chief qualifications.

If you do not have inclinations along these lines, it might be better to buy your equipment factory built. But if you have these temperament prerequisites, kit building can be a wonderful hobby. The quiet, concentrated routine of putting equipment together piece by piece can be a source of relaxation for many people.

Only about half a dozen very inexpensive tools are needed for kit building; these include a pair of long-nose pliers, large and small screwdriver, a wire cutter and stripper, a small adjustable wrench, and a very light soldering iron (Fig. 8-3).

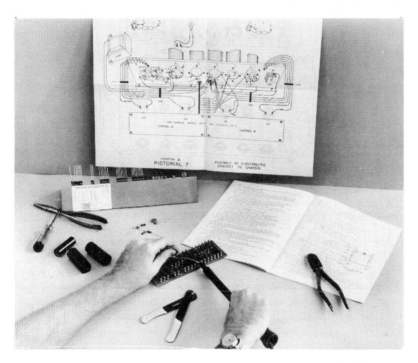

Fig. 8-3. Typical kit-builder's work area. Note tools, instruction book, and oversize diagram hung on wall.

Building a power amplifier is a simple kit project, for such amplifiers have relatively few parts and no intricate switching circuits. Building a preamplifier or an integrated amplifier involves more complicated wiring and hence a somewhat greater measure of dexterity. Tuners used to be the most difficult kits to build since the wiring layout of radio circuitry had to be kept to very close tolerances. In recent years tuner kits have been greatly simplified and almost nothing is left to chance. All critical parts are pre-assembled and adjusted at the factory; hence, you need have few qualms about tackling an fm tuner project.

CHAPTER 9

Setting Up *Your Sound System*

Hooking up a high-fidelity system is in principle almost as simple as plugging in a lamp. Certainly the mechanical aspect of it is hardly more complicated. The trick is to know just which connection goes where, and even that is no trick if you keep your mind on the basic logic of the system, visualizing how the audio signals travel from one component to another until they finally emerge from the speaker as audible sound (Fig. 9-1).

This logic is best derived by classifying components according to their function. First there are the program sources, the components that originate the signals you hear. The turntable and tone arm, tuner, and tape recorder or player fall within this group. The signals coming from these program sources are connected to the inputs of the amplifier, which serves as the center of the whole system.

In stereo, each incoming signal arrives at the amplifier by a pair of cables—one for the left channel and the other for the right. Often the left channel is marked Channel A, and the right channel is marked Channel B. It is helpful, though by no means necessary, to have the input cables color-coded so that at a glance you can differentiate between the cables belonging to the left and right channels. The corresponding input terminals at the amplifier will be clearly marked.

Assuming that you are using a magnetic cartridge in your tone arm (since most high-fidelity cartridges are magnetic), you take the small pointed pin plugs at the end of the record-player leads and shove them firmly into the round input openings on the amplifier that are marked either PHONO or MAG PHONO or sometimes simply MAG. Be sure these pin plugs

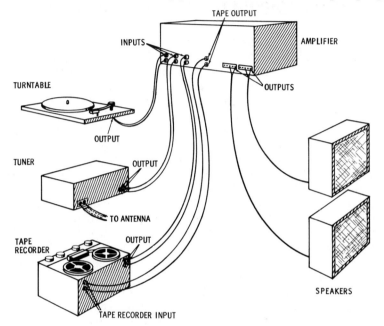

Fig. 9-1. Audio connections in a typical stereo system.

are seated firmly in their receptacles. Be sure they make contact at the outer rim, which carries the so-called ground connection. If that contact is loose, you get a condition known as an "open ground", and a loud hum results.

Sometimes a third cable without a pin plug at the end comes from the tone arm. This is an additional ground wire intended to minimize hum. Attach it to any convenient spot on the chassis of the amplifier—usually a chassis screw.

Next connect your tuner. The procedure is basically the same, except that usually there are no cables coming from the tuner. Instead, on the tuner you will find small annular terminals just like the input terminals on the amplifier but marked OUTPUT. You can buy a so-called audio cable (or patch cord) of the required length at your audio dealer. These cables have pin plugs at both ends. One end is inserted at the tuner, and the other end is inserted at what is marked as the tuner input on the amplifier. Again make sure you don't get the left and right channels mixed up. In four-channel systems, you also have to keep the front and rear connections apart.

SPEAKER HOOKUP—IMPEDANCE MATCHING AND PHASING

The next step is to connect the speakers to the amplifier. This must always be done *before* the amplifier is turned on, because it would damage the amplifier to run it if the speakers were not connected. This connection can be made with ordinary lamp cord, which you can buy in the required length at any hardware store. Strip about half an inch of insulation off the ends and tie one pair around the screw terminals found at the rear of the speakers. The other pair goes to your amplifier output terminals. These are usually two strips of screw terminals, one strip for each channel. The numerical markings alongside the screws denote the *impedance* of the terminal. Impedance is an electrical measure of electrical devices, including speakers. Various speakers have different impedances, usually 8 or 16 ohms, but some are rated at 4 ohms or other values. The hookup instructions furnished with your speakers will specify the proper impedance for your particular model. The wire coming from the terminal marked plus (+) on the rear of your speaker should go to the amplifier terminal marked with the correct impedance number. The other wire in the lamp cord, coming from the terminal marked minus (−) on your speaker should go to the amplifier output terminal marked COM, O, or sometimes G. These connections are made by means of a screwdriver. Before you clamp the wire behind the screw, twist the ends so that no loose strands stick out. If loose strands or wire ends accidentally touch the other wire or the neighboring screws, a short circuit might result that would impair the function of your speaker and might even cause damage to the amplifier.

In a stereo system it is vitally important that the two speakers operate "in phase." This means that their cones should be working in tandem—pushing and pulling at the same time rather than working against each other. If one pushes while the other pulls, their actions cancel each other and much sonic richness is lost. If your speakers have the mentioned plus and minus markings at their rear terminals (sometimes there is a COM marking instead of the minus sign), all you have to do is to follow the procedure just described, and your speakers will be in phase automatically. But if your speaker terminals are not keyed in this manner, you can assure proper phasing as follows: Throw the "phase reverse" switch on your amplifier back and forth and listen carefully to ascertain in which position the bass is fullest and the overall sound more smooth. Then leave the switch in that position. If you do not have a phase reverse switch, you can accomplish the same effect by reversing the wires at the rear of *one* of your speakers and observing whether this brings about an increase in bass and general smoothness of sound. Be sure you switch wires at one speaker only; if you do it on both speakers, you will wind up with exactly the same phase relations as before.

This is the basic hookup procedure. Of course, there are variations on this simple process, depending on the particular components employed in your system. For instance, if you are using a separate power amplifier and preamplifier, you need two cables (one for each channel) to connect the output of the preamplifier to the input of the power amplifier. Again this is done with audio cables equipped with pin terminals.

So far only the signal-carrying wires between the various components have been discussed. In addition to this, you must also attach the power cords by which the various components draw their electric supply. At the rear of most amplifiers (or preamplifier if you are using a separate preamp) you will find regular ac outlets similar to ordinary wall sockets. Connect the power plugs of your turntable, tuner, and tape recorder to these receptacles. This will provide the convenience of being able to turn all your components on and off with a single switch—the main power switch of your amplifier.

TURNTABLE ADJUSTMENTS

Before you play your first record, make sure that the stylus pressure of your tone arm is correctly set to the value recommended by the manufacturer of your cartridge. This check and any adjustment that may be necessary usually require a stylus pressure gauge (Fig. 9-2) obtainable at a nominal price from your audio dealer. Some tone arms have built-in calibrated scales for setting stylus pressure.

Next, put a small spirit level (Fig. 9-3) on your turntable. (These levels are available as an accessory in audio stores, but almost any small level will do.) This will show whether your turntable is tilted or not. Most turntables have screw adjustments by which they can be set level—something that must be done if your floor or furniture is out of plumb.

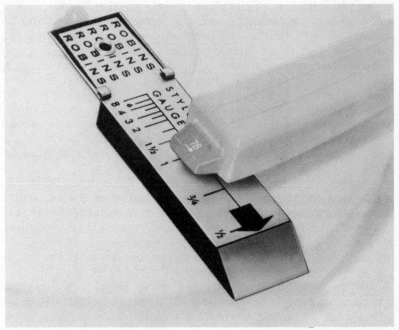

Courtesy Robins Industries Corp.

Fig. 9-2. A typical stylus-pressure gauge.

Of course these turntable adjustments must be made with the cartridge already mounted in the tone arm. Cartridge mounting is not intrinsically difficult, but it requires a certain amount of dexterity. The instructions furnished with the cartridge are usually explicit, but if you are not mechanically adept, you can easily injure the cartridge during the mounting process. For that reason it may be best to let your audio dealer install the cartridge. It's a good idea, too, to remove the vulner-

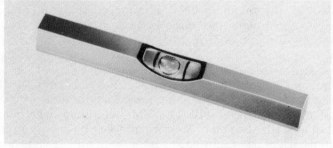

Courtesy Robins Industries Corp.

Fig. 9-3. A typical turntable level.

able stylus from the cartridge while you are transporting or installing your turntable. Whenever your record player is not in use, the tone arm should be firmly held in place so that it will not be accidentally knocked from its resting position. If the cartridge jars against the turntable, the stylus will probably be damaged. *When not in use, the stylus should never touch anything.*

INPUT LEVEL CONTROLS

A good many amplifiers and preamplifiers have so-called level sets (also known as input-level controls). As the name implies, these are auxiliary volume controls, usually located at the rear of the chassis near the input terminals. Their purpose is twofold: to avoid overloading the amplifier from an excessively strong signal source, and to let you adjust the relative volume of the various signal sources (record player, tuner, etc.) so that they will sound equally loud when you switch from one to the other. That way you do not have to readjust the main volume control when you change from records to tape or fm or vice versa.

The following procedure will help you set these controls to a satisfactory level. Set the main volume control of your amplifier approximately to the midpoint of its rotation. Then adjust the level sets at the rear to provide normal listening level on all the various inputs. Once you have set these controls, you can forget them.

ACOUSTIC FEEDBACK

The turntable should not be positioned on the same shelf or the same piece of furniture as the speakers. Otherwise the vibrations from the speakers may travel back to the turntable, causing a condition known as *acoustic feedback*. The pickup then repeats the vibrations that have already gone through to the speakers and are now coming back. This results in a tonal blur, a low grumbling, a thumping noise, and sometimes a "banshee" howl. This may happen if the turntable and speakers are not close together and the speaker vibrations travel back to the turntable through the floor boards. Small foam-rubber pads placed under the speakers to soak up excess vibration are a handy remedy. If that is insufficient, make the turntable mounting more elastic by putting a foam-rubber pad under the turntable base or moving the turntable to some less shaky location.

VENTILATION

Overheating is the chief cause of equipment failure. Giving your amplifier and tuner ample ventilation for cooling may lengthen their life by years. Since hot air rises from the equipment, you should never stack components on top of one another. Don't place books, magazines, or any other objects that will inhibit free heat dispersion on your components. Make sure there is at least a 6-inch air space above each component, and at least a 2-inch free-air space between the rear of the chassis and the nearest wall.

If your equipment sits on open shelves, you will not normally encounter any problem of overheating. However, if the equipment is built into cabinets, it is necessary to assure free air flow around the components. Vents should be provided near the bottom and near the top (usually at the rear) so that a forced draft results, with cool air being taken in at the bottom and warm air escaping at the top. Again, a gap should exist between the cabinet and the wall behind it so that this air flow will not be obstructed.

If your equipment is housed in cramped quarters, it may be advisable to install a small cooling fan such as the one illustrated in Fig. 9-4. Small, silent fans, designed especially for this purpose, are available at the larger audio shops. If you are in doubt about overheating, consult a competent serviceman.

Courtesy Rotron, Inc.

Fig. 9-4. A small, silent fan designed for cooling audio power amplifiers mounted in unvented cabinets.

CHAPTER 10

Getting the Most from Your Sound System

To get the best performance of which your sound system is capable, you should know the function of its various controls, place the speakers in an acoustically favorable spot, keep your records in good condition, and provide the tuner with a suitable antenna.

CONTROLS

The purpose of the volume control is self-explanatory. Some amplifiers have two separate volume controls—one for the left channel and the other for the right, usually arranged concentrically. By means of these separate controls you can balance the volume of the two channels until both seem equally loud from where you sit. On other amplifiers this is accomplished by a single volume control plus a separate balance control. You set the volume control for the loudness you like and then use the balance control to even out the volume between the two speakers until they both sound equally loud.

Another control is the input selector by which you choose the program source you want to hear: phono, tuner, or tape. This control is often combined with a so-called mode selector, though on some amplifiers the mode selector is a separate knob. The

97

main purpose of the mode selector is to choose between mono and stereo operation. The mono setting combines Channel A with Channel B. On some equipment you have the added option of feeding either channel separately to both amplifier outputs. Other selector functions are often provided.

Tone Controls

Treble and bass controls, commonly called tone controls, fulfill the obvious function of either boosting or attenuating the treble and bass frequencies. In their neutral position (usually pointing straight up) they neither boost nor attenuate any frequencies but provide flat frequency response. For most records and broadcasts this "flat" position results in natural tonal balance. For this reason some listeners hesitate to move these controls from their normal setting; however, the tonal flexibility they provide can contribute much to your musical enjoyment.

One important use of tone controls is to compensate for variations in the records themselves. Records differ widely in tonal character—partly because of the different acoustic environments in which the original performances took place, different microphone placement at the various recording sessions, and different engineering practices at the various record companies. Thus, some records sound bright and brilliant, while others sound warm and rich. The tone controls let you compensate to some degree for these differences and attain a tonal balance in keeping with your personal preference. Many records do not contain sufficient bass to render the feeling of orchestral mass and fullness. The reason for this is that the powerful bass amplitudes are difficult for the stylus to track on ordinary phonographs without good tone arms. Some record companies therefore make it a policy to weaken the bass somewhat so that average equipment will encounter no trouble. The bass control can be adjusted to compensate for this. Treble adjustments are also often needed to achieve tonal balance that is pleasing to the listener.

In addition to the regular tone controls, many amplifiers also have a control called "loudness compensation." Its purpose is to retain a sense of tonal fullness in music played at low volume. One of the peculiarities of human hearing is that we do not hear soft sounds in the same way we hear sounds at higher loudness levels. Specifically, human hearing does not register low notes at low levels. In consequence, if music is played at a lower volume than it would be naturally heard in a concert, the bass seems to drop out, making the sound appear thin and tinny. Loudness compensation is used to boost the bass approximately the same amount that the human ear suppresses it at low volume. So, subjectively, the impression of the music stays about the same, regardless of the volume level at which you play it. Your sense of the music's natural depth and richness thus remains unaltered. (In addition to bass boost, loudness compensation also provides a lesser correction of high-frequency response to make up for the changed character of human hearing at low volume in that range.)

Blend Control

The blend control (also called separation control) provided on many stereo amplifiers and some stereo tuners lets you cross-feed varying amounts of signal from Channel A to Channel B, and vice versa. This regulates the amount of stereo separation. If your speakers are too far apart (possibly for reasons of convenience or decor), the sound sources, too, will seem too far apart with not enough sound seeming to emerge from midway between the speakers. This is known as the hole-in-the-middle effect. The blend control corrects this condition. When you select a judicious amount of signal cross-feeding between the two channels, you compensate electronically for the extra distance between the speakers and restore the unbroken spread of the apparent sound source.

Other Controls

Other controls you may find on your set, such as scratch and rumble filters, are mostly self-explanatory. It merely remains to point out that such filters should never be switched on unless they are really needed because of scratchy records or rumbling turntables. They operate by cutting off the top and bottom of the frequency spectrum, partly negating the advantage of high fidelity.

Controls for four-channel systems are basically similar to those for standard stereo. However, there is an additional control for adjusting the balance between front and rear, as well as a control for the side-to-side balance of the two rear channels. The master volume control on "quad" systems acts simultaneously on all four channels and thus maintains the set balance between the four speakers regardless of volume.

ROOM ACOUSTICS AND YOUR FAVORITE CHAIR

Unless you have experimented with speaker placement, you would hardly believe the enormous difference that can result from the location of the speakers in a given room. Your room, with whatever acoustic quirks it may have, is as much a component of your sound system as any of the equipment you buy. Technically speaking the room is the "acoustic load" into which the speakers deliver their output. That is why the placement of speakers with respect to the shape of the room is of vital importance for the effectiveness of sound projection (Fig. 10-1).

Since no two living rooms are quite alike in size, shape, and furniture arrangement, no universally applicable rules for speaker placement can be given. But a few pointers might help you find more quickly the kind of speaker placement that brings the best results in your home.

In general, stereo speakers are placed from 8 to 12 feet apart in average-size rooms. The optimum listening position is across the room from the speakers, somewhere close to the midpoint between the speakers. However, this pattern need not be observed strictly. Contrary to popular belief, you need not sit right in the middle between the speakers. Stereo allows the listener much more freedom in choosing a convenient listening spot than is generally assumed, so don't feel hemmed in by the geometry of stereo. You can get all of the fullness and spaciousness of stereo as well as much of its left/right directionality even if your favorite listening chair isn't located right at the apex of an equilateral triangle whose base is defined by the two speakers. The balance control on your amplifier permits compensation for off-center seating.

You get better bass projection by placing your speaker or speakers in the corners as shown in Fig. 10-2. It also increases the bass to let the speakers face the whole length of the room rather than across the shorter width of the room. For maximum bass projection the room dimension faced by the speakers should be at least 15 to 17 feet. Smaller rooms present an inherent handicap to full bass reproduction.

You usually get better bass response if you set your speakers down on the floor. However, many people find it unnatural to have the sound source so low and prefer raising their speakers about five to seven feet off the floor so that the sound comes from slightly above ear level.

By all means experiment with unorthodox speaker placements. For instance, you can put the speakers more than the usual distance apart and adjust for the added physical separation by means of the blend control. Or you can put the speakers at angles, looking diagonally into the room. You may even have them set at right angles along adjoining walls and put your listening chair near the intersection of their two lines of sound projection.

Some speakers have controls at their rear to adjust the relative loudness of the tweeter with respect to the woofer. When

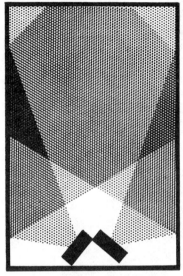

(A) Speakers placed so that sound reflects from the walls of the room.

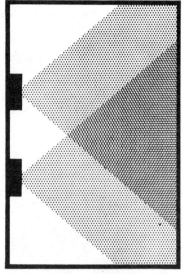

(B) Speakers placed to project toward the short dimension of the room.

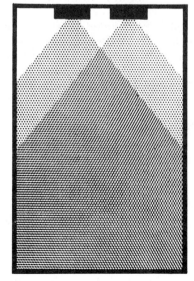

(C) Speakers placed to project toward the long dimension of the room.

Fig. 10-1. Three different speaker arrangements.

Courtesy University Sound

Fig. 10-2. Placement of speaker in corner to improve bass response.

you have decided on the proper location of your speakers, ask someone to turn these tweeter controls slowly through their range while you sit in your usual listening place. You will then be able to determine the tweeter-level setting you like best. For this test play a first-rate record with plenty of highs; play it at normal volume, with treble and bass in a neutral position.

Incidentally, if your speakers have no tweeter level control, this is by no means a reflection on their merit. If you feel the need of treble adjustment, you can get approximately the same effect without tweeter controls by means of the treble controls on your amplifier.

Four-Channel Speaker Placement

For optimum four-channel sound, the listener should sit at a distance approximately equal from all four speakers. This could be attained by putting the speakers in the four corners of the room and the listener somewhere near the center (Fig. 10-3A). Such an arrangement is clearly impractical in most

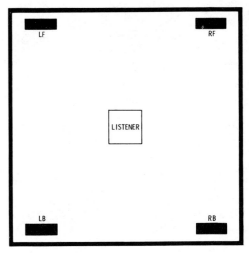

(A) Listener in center of room.

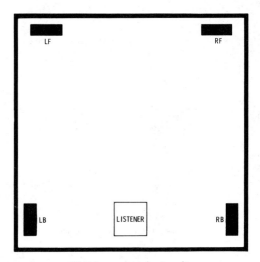

(B) Listener along back wall.

Fig. 10-3. Four-channel speaker placement.

living-room settings, where the listener usually sits on a chair or couch placed alongside one of the walls. To get a good four-channel effect in such typical living-room situations, place the rear speakers in the corners adjacent to the wall where you sit while facing the front speakers across the room (Fig. 10-3B). Then adjust the front-to-rear balance so that most of the sound comes from the front while the rear speakers chime in more softly. The exact proportion of front-to-rear sound is a matter of individual taste, but in no case should the rear speakers overbalance the sound from the front.

KEEP IT CLEAN

Remember that your components do not originate the audio signals you hear. They only reproduce the signals given to them, and if those original sound sources are tainted with noise and distortion, not even the best components can improve them. On the contrary, the better your system, the more faithfully it will render not only the music but also the defects in the original signal. To get the best performance of which your components are capable, you must keep the signals "clean."

For fm tuners this means that you must provide them with an antenna adequate to the needs of your particular receiving location. The types of antennas suitable for various typical receiving conditions have already been mentioned under tuner section. As the stylus tip slams into the dust grain with this enormous force, the grain (which is essentially a piece of rock with jagged cutting edges) is driven into the soft vinyl record material and does permanent damage (Fig. 10-4). No later amount of cleaning can undo this damage.

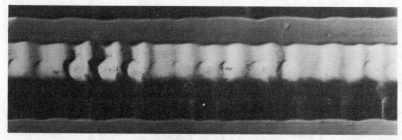

(A) Groove worn by defective stylus.

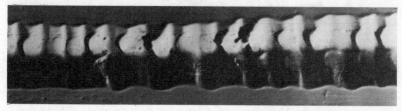

Courtesy Fabric Research Laboratories, Inc.
(B) Groove pitted by dust under stylus.

Fig. 10-4. Magnified views of damaged record grooves.

Try to make a habit of cleaning your records before every play. After playing, once the dust is dug in, it's too late. Various record-cleaning aids are available at audio shops and at better record stores. Velvet cleaning pads do very nicely, because the nap of the fabric gets down into the grooves. A soft camel's-hair brush about three to five inches wide (the kind used by photographers to clean their negatives) is also helpful. Another convenient record-cleaning device is the "dust bug," which rides the record ahead of the tone arm and sweeps the grooves before the stylus plays them. The "dust bug" is shown in Fig. 10-5.

Courtesy Elpa Marketing Industries, Inc.

Fig. 10-5. Convenient record cleaning device.

Stick to a hands-off policy as far as your records are concerned. Your oily fingertips are perfect dust catchers. Learn how to handle your records by touching them only at rim and label. When you slide records from and into their jackets, squeeze the jacket sides outward to prevent chafing the record surface. Store your records standing up vertically. Do not stack them so that there is pressure on the sides. Don't leave your discs lying about without their protective sleeves. Put them back in their jackets immediately after playing. Of course, there is no point in putting a clean record on a dirty turntable, so keep a dust cover on your turntable when it is not in use, and occasionally sweep the mat on which the records rest.

If, despite all these precautions, your records have collected dust, put them through a record "laundry" with cold water, a weak detergent solution, and a cellulose sponge. Then dry them with a lint-free Turkish towel.

Quite possibly, this cleanliness routine may seem rather hypochondriacal at first glance. But remember that you are not only protecting your cash investment in records by keeping them clean, but also the priceless musical values on irreplaceable discs.

TAPE CARE

It is not generally realized that tapes, like records, also require a certain amount of care if they are to last for a long time and remain at their best.

The first step in assuring long life for your tapes is to pick the right sort. Acetate-based tape, for example, has an inherently limited life span because its plasticizer—the chemical to make the tape pliant—gradually evaporates. After about ten years, such tapes often become too brittle to be played, even if they are stored in sealed boxes. Moreover, acetate is affected by temperature and humidity, which cause it alternately to stretch and contract. As a result, the tape crinkles and buckles on the reel, and when it is played again, it weaves erratically about the playback head, producing dull, wobbly, or intermittent sound. The main reason acetate tapes are favored for professional applications is that they are less likely to be damaged in editing.

Far superior in storage characteristics are tapes based on polyester (popularly known under the trade name *Mylar*) or tapes made with polyvinyl chloride (PVC). These compounds are chemically stable and retain their strength and pliancy indefinitely.

Proper handling of tapes is a key factor in keeping them in good condition. The following hints will help avoid some of the common causes of tape damage:

1. Rewind tapes before playing, not after. The fast rewind speed of the tape recorder puts too much tension on tape for long-term storage. Tape should therefore be stored on the take-up reel after winding up slowly during recording or playing.
2. Store tape boxes vertically on their edges. Letting them lie flat makes the tape sag on the reel, causing poor alignment during playback.
3. Keep the tape recorder clean. Dust and oxide particles on the head and tape guides abrade the tape. Avoid frequent starting and stopping during playback to keep tape tension even. Have tension on recorder adjusted if starts and stops are jerky.
4. Keep tape away from heat and direct sunlight as well as from magnetic fields around speakers and transformers.

Even an ordinary ac lamp cord can cause magnetic trouble if tapes are left near it for long periods.
5. Check tape reel for even wind. Rewind reel slowly if the edges of individual turns stick out from the pack.

When these points are observed, tapes will outlast any other recording medium without loss of fidelity.

HOW LOUD SHOULD YOU PLAY IT?

Fidelity to the musical original is obtained when you play music at the same volume level at which it would be heard in a concert. This doesn't mean that your system has to put out as much sound as an actual orchestra because, after all, your living room is smaller than a concert hall. The point is that the sound of a given instrument should reach you with approximately the same loudness in your chair at home as it does in your concert seat—and this doesn't necessarily mean the front row.

CHAPTER 11

First Aid

As with any kind of machinery, you may occasionally encounter malfunctions in your sound system. Since components are usually very reliable and built for years of trouble-free service, such difficulties are often minor, and you may be able to cure them without resorting to professional service. The important thing is to know the symptoms that help you track down the trouble.

If the entire system or one of its components simply goes dead and won't even light up, check the power plugs; they may have come loose. Then check the fuses mounted on the rear of the amplifier and tuner. Replace them with fuses of identical ratings if they are burned out. If the new fuse also blows, it's a sign of interior trouble, and you should let a competent serviceman take over.

HUM

Probably the most common hi-fi trouble is hum. Usually hum is heard only on one program source, either the tuner or the record player. The most likely cause is a loose plug on a connecting cable. Shove the pin plugs on the input leads from the "hummy" component firmly into their sockets, pressing down on their outer shields and twisting them slightly to scrape off any corrosion that may have formed on the contact surfaces. If this doesn't help, check the input cables for breaks around the pin terminals (Fig. 11-1), or replace the cable. If the hum occurs on the phono channels, check the little pin jacks on the cartridge terminals. Squeeze them tight and move them slightly with a pair of tweezers to break down any film of corrosion that may have formed between the contact sleeve and the cartridge terminal pins. Follow the same procedure if one of the channels goes dead. This condition, too, may be caused by loose connections between the program source and the amplifier.

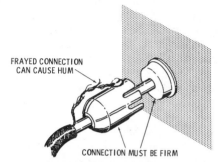

Fig. 11-1. Hum can be caused by loose plug connections.

STYLUS PROBLEMS

Distortion may have many causes, but if it occurs only on the phono input, chances are that the trouble lies with the stylus. It may have been accidentally bent, or it might have become clogged with accumulated grime from the grooves. A bent stylus must be replaced. A dirt-encrusted stylus can be cleaned with a stylus brush, or the dirt accumulation may have to be pried from the stylus with the tip of a pin. This job must be performed with the utmost delicacy to prevent damage to the stylus.

It is possible, of course, that your stylus may simply be worn, though a diamond stylus in a compliant cartridge played at light pressure should last several years in normal use. Make it a practice to take your cartridge at least once yearly to your audio dealer for a quick microscopic checkup.

A defective stylus or improper stylus pressure may also cause the tone arm to skip grooves. It's a good idea to check stylus pressure from time to time with a stylus pressure gauge. If necessary, reset the stylus pressure to the value recommended by the cartridge manufacturer. Also make sure the turntable isn't tilted; a small spirit level is handy for this purpose.

OTHER PROBLEMS

If fm reception suddenly deteriorates and you get only the strongest stations, it may be that the antenna leads have come loose somewhere. Check the antenna connections at the rear of the tuner. Then check the antenna leads all the way back to the antenna itself.

If the speaker conks out, the trouble may simply be a shorted speaker lead. See if a pair of wires are accidentally touching at the screw terminals on the amplifier or at the rear of the speaker. If this isn't the ready answer to your difficulty, connect the dead speaker to the other amplifier channel and see if it revives. If so, you know that one channel in your amplifier is dead—off it goes to the service shop.

What may appear to be a rattle in the speaker is often caused by nearby window panes, lamps, crockery, and bric-a-brac. Blame the speaker only after having eliminated these likely possibilities. What seems to be a speaker rattle is often merely evidence of poor stylus tracking on the record. You know this is the case if the rattle ceases when you play the tuner. If so, check stylus condition and tracking pressure.

Many sudden hi-fi troubles are purely external to the components. A simple check of interconnecting cables, as outlined here, will disclose and dispose of most common complaints. But in case the difficulty lies in the electronic workings of your equipment, the first thing to do is to isolate the component at fault. If all program sources work except the tuner, it is obviously the tuner that needs attention. If the difficulty occurs on all the various program sources, it is the amplifier that is defective. Determine also whether the difficulty occurs in both channels or only in one. This further helps isolate the source of trouble.

When simple home remedies fail, take the faulty component to a service shop. The larger high-fidelity dealers usually maintain competent service departments or can at least refer you to a trustworthy service firm. You may inquire from the manufacturer of your equipment about authorized service facilities nearest to you. Many tv and radio shops are also able to service high-fidelity equipment.

There is no need to be apprehensive of serious difficulty. If you use your equipment according to instructions, install it properly and treat it considerately, it should give you years of trouble-free listening.

Index

A
Accuracy, speed, 63-64
Acoustic feedback, 95
Acoustics, room, 99-103
Adjustments, turntable, 93-95
Afc, 46
Ambience, 16
Amplifiers(s), 34-42
  and preamplifier, 29-30
  power rating, 34-37
Amplitude, 11

B
Bantam speakers, 54-55
Bass response, 49-50
Blend control, 99
"Bottom dollar" stereo, 77

C
Capture ratio, 45
Cartridge, 29
  types, 59-60
CD-4, 18-19
Compact systems, 25-26
Comparison of systems, 24-25
Compliance, 55-56
Component systems, 23-24
Consoles, 21-23
Control
  blend, 99
  input level, 95
  tone, 98-99
Cost factor, 17-18
Cycle, 8

D
Decibel, 40
Deluxe stereo, 79-81
Dimensions, stylus, 58
Directionality, 50-51
Distortion, 37-38
  harmonic, 38
  intermodulation, 38
Drag, friction, 65
Dynamic mass, 57

E
Earphones, 73-75
Effect, stereo, 12-13
Enclosures, 51
Error, tracking, 64-65

F
Feedback, acoustic, 95
Flutter, 27
Four-channel budgeting, 81
Frequency
  of a tone, 8
  ranges, 10
  response
    and resonance, 58-59
    flat, 39
Friction drag, 65
Fundamental, 9

G
"Golden medium" stereo, 77-78

H
Harmonic distortion, 38
Harmonics, 9
Hertz, 9
Hookup, speaker, 92-93
Hum, 108-109
  and noise, 42
Hysteresis-synchronous motors, 63

I

IHF standard, 37
Input level controls, 95
Intermodulation distortion, 38
Inverse feedback, 38

K

Kit assembly, 86-88

L

Listener fatigue, 35
Loudness, 10-11

M

Mass, dynamic, 57
Matrixing, 18
Modulation, 11
Motors, hysteresis synchronous, 63
Multiplex, 32

O

Open reel machines, 69-71
Operating features, tuner, 46
Overtones, 9

P

Phasing, speaker, 92-93
Phonocartridges, 55-60
Pickups, 29
Pitch, 8-9
Power rating of amplifiers, 34-37
Pressure, tracking, 57, 65

Q

Quad
  equipment, 19-20
  program sources, 18-19
Quadraphonic sound, 16
Quieting, 44

R

Receiver, 32-33
Record changer, 28-29, 68
Resonance, 66
Response
  frequency, 39-40
  transient, 40-42
Rms method, 37
Room acoustics, 99-103
Rumble, 28, 60-62

S

Separation, stereo, 59
Skating compensation, 67-68
Sound
  and electricity, 11
  quadraphonic, 16

Sources of quad programs, 18-19
Speaker, 30-31, 47-54
  hookup, 92-93
  phasing, 92-93
Specifications, tuner, 43-46
Speed accuracy, 63-64
SQ matrix, 18-19
Squelch circuits, 46
Standard, IHF, 37
Stereo
  "bottom dollar," 77
  deluxe, 79-81
  effect, 12-13
  "golden medium," 77-78
  reproduction, 13-15
  separation, 59
Stylus
  dimensions, 58
  problems, 109-110
Systems
  compact, 25-26
  comparison of, 24-25
  component, 23-24

T

Tape
  cassettes and cartridges, 71-72
  care, 106-107
  recorders, 68-73
  open reel, 69-71
Timbre, 9
Tone
  arm, 28, 64-68
  color, 9-10
  control, 98-99
Tracking
  error, 64-65
  pressure, 57, 65
Transient response, 40-42, 53-54
Treble response, 49
Tuner, 31-32, 42-47
  operating features, 46
  sensitivity, 43-45
  specifications, 43-46
Turntable, 27-38, 60-64
  adjustments, 93-95
Tweeter, 30
Types of tuners, 43

V

Ventilation, 96

W

Woofer, 30
Wow, 27
Wow and flutter, 62-63